Host Cell Capture-Quantitative Sequence Detection of Potentially Infectious Viruses

Awwa Research Foundation

About the Awwa Research Foundation

The Awwa Research Foundation (AwwaRF) is a member-supported, international, nonprofit organization that sponsors research to enable water utilities, public health agencies, and other professionals to provide safe and affordable drinking water to consumers.

The Foundation's mission is to advance the science of water to improve the quality of life. To achieve this mission, the Foundation sponsors studies on all aspects of drinking water, including supply and resources, treatment, monitoring and analysis, distribution, management, and health effects. Funding for research is provided primarily by subscription payments from approximately 1,000 utilities, consulting firms, and manufacturers in North America and abroad. Additional funding comes from collaborative partnerships with other national and international organizations, allowing for resources to be leveraged, expertise to be shared, and broad-based knowledge to be developed and disseminated. Government funding serves as a third source of research dollars.

From its headquarters in Denver, Colorado, the Foundation's staff directs and supports the efforts of more than 800 volunteers who serve on the board of trustees and various committees. These volunteers represent many facets of the water industry, and contribute their expertise to select and monitor research studies that benefit the entire drinking water community.

The results of research are disseminated through a number of channels, including reports, the Web site, conferences, and periodicals.

For subscribers, the Foundation serves as a cooperative program in which water suppliers unite to pool their resources. By applying Foundation research findings, these water suppliers can save substantial costs and stay on the leading edge of drinking water science and technology. Since its inception, AwwaRF has supplied the water community with more than $300 million in applied research.

More information about the Foundation and how to become a subscriber is available on the Web at **www.awwarf.org.**

Host Cell Capture-Quantitative Sequence Detection of Potentially Infectious Viruses

Prepared by:
George D. Di Giovanni
Texas AgriLife Research
Texas A&M University System, El Paso, Texas 79927
and

Kristina D. Mena and **Laura Y. Sifuentes**
University of Texas School of Public Health
El Paso Regional Campus, El Paso, Texas 79902

Sponsored by:
Awwa Research Foundation
6666 West Quincy Avenue, Denver, CO 80235-3098

Published by:

Awwa Research Foundation

Distributed by:

American Water Works Association

IWA Publishing

DISCLAIMER

This study was funded by the Awwa Research Foundation (AwwaRF). AwwaRF assumes no responsibility for the content of the research study reported in this publication or for the opinions or statements of fact expressed in the report. The mention of trade names for commercial products does not represent or imply the approval or endorsement of AwwaRF. This report is presented solely for informational purposes.

CONTENTS

TABLES

FIGURES

FOREWORD

The Awwa Research Foundation is a nonprofit corporation that is dedicated to the implementation of a research effort to help utilities respond to regulatory requirements and traditional high-priority concerns of the industry. The research agenda is developed through a process of consultation with subscribers and drinking water professionals. Under the umbrella of a Strategic Research Plan, the Research Advisory Council prioritizes the suggested projects based upon current and future needs, applicability, and past work; the recommendations are forwarded to the Board of Trustees for final selection. The foundation also sponsors research projects through the unsolicited proposal process; the Collaborative Research, Research Applications, and Tailored Collaboration programs; and various joint research efforts with organizations such as the U.S. Environmental Protection Agency, the U.S. Bureau of Reclamation, and the Association of California Water Agencies.

This publication is a result of one of these sponsored studies, and it is hoped that its findings will be applied in communities throughout the world. The following report serves not only as a means of communicating the results of the water industry's centralized research program but also as a tool to enlist the further support of the nonmember utilities and individuals.

Projects are managed closely from their inception to the final report by the Foundation's staff and large cadre of volunteers who willingly contribute their time and expertise. The Foundation serves a planning and management function and awards contracts to other institutions such as water utilities, universities, and engineering firms. The funding for this research effort comes primarily from the Subscription Program, through which water utilities subscribe to the research program and make an annual payment proportionate to the volume of water they deliver and consultants and manufacturers subscribe based on their annual billings. The program offers a cost-effective and fair method for funding research in the public interest.

A broad spectrum of water supply issues is addressed by the Foundation's research agenda: resources, treatment and operations, distribution and storage, water quality and analysis, toxicology, economics, and management. The ultimate purpose of coordinated effort is to assist water suppliers to provide the highest possible quality of water economically and reliably. The true benefits are realized when the results are implemented at the utility level. The Foundation's trustees are pleased to offer this publication as a contribution toward that end.

David E. Rager
Chair, Board of Trustees
Awwa Research Foundation

Robert C. Renner, P.E.
Executive Director
Awwa Research Foundation

ACKNOWLEDGMENTS

The authors would like to thank the Awwa Research Foundation for funding and support. This project was also funded by Texas AgriLife Research, Texas A&M University System. We are thankful for the comments provided by the Project Advisory Committee – John Chandler, Jan Vinje and Kevin Oshima. The support of the AwwaRF Project Officer, Misha Hasan, was also greatly appreciated.

The majority of the laboratory analyses were performed by Laura Sifuentes as part of her Master's of Public Health thesis. We also thank Michelle Sims and Joe Hernandez for technical assistance.

EXECUTIVE SUMMARY

Waterborne disease outbreaks associated with infectious agents continue to occur despite the implementation of advanced water treatment technologies with more than 700 outbreaks documented in the U.S. between 1971 and 2002. Further, some experts believe that many more outbreaks occurred that were not reported or recognized. Approximately 8% of these documented waterborne outbreaks were of viral etiology. However, based on the human health outcomes observed, it is speculated that almost half of those outbreaks for which the causative agent was never identified were likely due to enteric viruses.

The enteric viruses include the enteroviruses (poliovirus, coxsackieviruses, and echoviruses), rotaviruses, hepatitis A and E, Norwalk and Norwalk-like viruses ("noroviruses"), adenoviruses, reoviruses, and others. Enteric viruses are routinely detected in raw sewage, surface waters, and groundwater, are able to survive in the environment for long periods of time. Compared to bacteria, enteric viruses are more resistant to conventional water and wastewater treatment. In addition, they have the greatest infectivity among waterborne pathogens and subsequent outbreaks result in more illness cases than outbreaks due to other agents. Furthermore, enteric viruses are associated with a greater number of serious illnesses than any other group of waterborne pathogens. Enteric viruses are capable of causing gastrointestinal illness, respiratory disease, eye infections, aseptic meningitis, hepatitis, and more chronic health consequences like myocarditis and insulin-dependent diabetes mellitus. Endemic transmission of waterborne illnesses is estimated to contribute to as many as 12 million cases each year in developed countries. The immunocompromised populations such as the elderly, young children, HIV patients, chemotherapy patients, and pregnant women, experience the most severe health consequences. Due to the potential public health impact associated with exposure to even low levels of enteric viruses, a useful method for the routine monitoring of infectious (or potentially infectious) viruses in water is essential for risk assessment and protection of public health.

Conventional detection of infectious waterborne viruses involves the use of mammalian cell culture. However, there are many disadvantages of standard cell culture for the detection of viruses, including cost, two weeks or longer for completion of the assay, and generic detection of viruses. Many different polymerase chain reaction (PCR) methods for the detection of viruses have been developed, with the advantages of specific virus detection, speed (same day assays), and reduced cost. Despite these strengths, PCR methods for virus detection do not determine virus infectivity. In recent years, integrated cell culture and PCR (ICC-PCR) methods have attempted to bridge the gap between the two detection strategies. This approach typically employs a shortened period (e.g. 5 h to 3 days) of cell culture with replicating viruses detected by PCR. While this approach overcomes some of the limitations of cell culture or PCR alone, it is still more time-consuming than desired and poses challenges for virus quantitation. A method that could be incorporated into current virus concentration, cell culture and PCR-based detection strategies would facilitate routine monitoring for infectious viruses.

RESEARCH OBJECTIVES

The main objective of this research was to further develop a method referred to as host cell capture quantitative sequence detection (HCC-QSD) for the rapid detection of potentially infectious viruses in water. Specific objectives of this research include the following:

1. Evaluate different host cell lines and capture conditions for their ability to capture enteric viruses.
2. Evaluate standard and advanced nucleic acid extraction and purification methods for separating viral nucleic acids from host cell monolayer nucleic acids.
3. Determine the applicability of previously published PCR primer sets and TaqMan probes for enteroviruses (e.g. pan-enteroviruses) and mammalian reovirus for use with QSD.
4. Evaluate the ability of HCC-QSD to distinguish potentially infectious viruses from those inactivated by free chlorine and ultraviolet light (UV).

APPROACH

Attachment to host cells and intact nucleic acid are two requirements for viral infection and formed the basis of the HCC-QSD detection strategy. For the host cell capture (HCC) step, cell culture monolayers were incubated with virus samples for three hours or less to "capture" viruses, followed by washing to remove non-adsorbed viruses and sample debris. RNA from "captured" viruses was then extracted using a commercial RNA purification kit and quantified using QSD. These conditions, combined with the specificity of molecular detection, were used to selectively recover, detect, and quantify potentially infectious viruses.

Different cell lines, viruses and capture conditions were evaluated to further develop the HCC step. Buffalo Green monkey kidney (BGMK) and Vero cell lines were evaluated for their ability to capture attenuated poliovirus type 3 (Fox strain) and mammalian reovirus type 3 (Dearing strain). Different sample and cell monolayer contact times, including less than 1 minute, 30 minutes, and 1-3 hours were evaluated for the optimization of the HCC step. In addition, different volumes of inocula and types of inoculation media were compared to determine the conditions most favorable for maintenance of intact monolayers and capture of viruses.

Nucleic acid extraction and purification experiments were aimed at reducing QSD amplification interference by host cell monolayer nucleic acid. The commercial Qiagen RNeasy and Viral RNA extraction kits were tested for their compatibility with the HCC-QSD method. The effect of various kit components on QSD amplification were evaluated, including the effect of carrier RNA for the Viral RNA kit and the use of QIAshredders and addition of β-mercaptoethanol to the lysis buffer for the RNeasy kit. Oligo capture immunomagnetic separation (IMS) was also investigated for the selective separation of viral RNA from cell monolayer RNA. DNase treatment was utilized as an addition to the RNeasy extraction kit, and was evaluated for its effectiveness in reducing PCR interference from DNA co-extracted from cell monolayers. Finally, different volumes of cell monolayer lysates and purified RNA used as PCR template were evaluated for their effect on HCC-QSD.

Inactivation studies to assess the ability of HCC-QSD to distinguish between infectious and disinfected (inactivated) viruses were performed using the model system of poliovirus 3 and BGMK cells. Poliovirus was exposed to zero, low, medium and high doses of low-pressure ultraviolet light (UV) and concentration × time (CT) levels of free chlorine. For these experiments, samples were analyzed using both HCC-QSD and conventional plaque assay.

CONCLUSIONS

Significant progress was made towards the development of HCC-QSD as a rapid method for the detection and quantitation of potentially infectious viruses in water. Specific conclusions of the project are:

1. Although the BGMK and Vero cell lines are best suited for the propagation of poliovirus and reovirus, respectively, either cell line could be used to capture the viruses. This provides some evidence for the possible use of a single cell monolayer for HCC-QSD to detect different viruses in the same sample. However, the capture dynamics were different between the two cell lines and viruses, indicating that further optimization would be required for this approach.
2. Monolayers grown in 12-well plates and inoculated with 100 μl sample volumes proved to be effective for the virus capture step.
3. The preparation of inocula in serum-free Minimum Essential Medium (MEM) is recommended over the use of phosphate buffered saline (PBS). The use of MEM helped maintain cell monolayer integrity compared to PBS, especially for the BGMK cell line.
4. Comparison of the two Qiagen RNA extraction kits revealed that the RNeasy kit without the use of QIAshredders or β-mercaptoethanol provided the most reliable extraction of viral RNA from cell monolayers.
5. Oligo capture immunomagnetic separation (IMS) was capable of separating viral RNA from host cell monolayer nucleic acids. However, RNA recovery efficiency was unacceptably low and further optimization of oligo capture conditions are needed prior to using this purification approach.
6. Addition of DNase treatment to the Qiagen RNeasy extraction protocol was shown to have a significant and beneficial effect on QSD amplification. The benefit of DNase treatment was especially pronounced with the BGMK cell line, and was reflected not only in increased QSD sensitivity, but also amplicon quality as detected by conventional gel electrophoresis.
7. In contrast to the conventional plaque assay, HCC-QSD was only capable of detecting UV-inactivation of poliovirus at doses of $\geq$ 40 mJ cm^{-2}. The greater sensitivity of the plaque assay compared to HCC-QSD was likely due to the inability of UV-inactivated virus to replicate in cell culture, whereas significant UV-induced damage to the viral genome was required before a decrease in HCC-QSD signal was observed. Therefore, HCC-QSD is currently not recommended for UV-inactivation work.
8. There was excellent agreement between HCC-QSD and the conventional plaque assay for the detection of infectious and free-chlorine inactivated virus. In repeat experiments both HCC-QSD and the conventional plaque assay detected complete inactivation of poliovirus with 0.5 mg·min L^{-1} free chlorine concentration $\times$ time (CT). This suggests that the HCC-QSD assay may be a useful alternative to the plaque assay for the rapid monitoring of chlorine-disinfected water.

FUTURE RESEARCH

Further development of the HCC-QSD method is needed prior to routine field application. In addition to poliovirus and reovirus, we had intended to work with hepatitis A and adenovirus 40 and 41 for this project. However, we were unsuccessful in propagating these viruses. Therefore, these viruses as well as others should be evaluated for use in HCC-QSD. Chlorine disinfection studies with these other viruses should be performed to understand the strengths and limitations of HCC-QSD.

Since HCC-QSD is potentially capable of recovering a variety of enteric viruses from a single sample, multiplex QSD primers and probes are needed to allow simultaneous detection of different viruses. With the current availability of novel PCR reagents to overcome some of the recognized problems with multiplexing, the use of previously developed primers and probes that are compatible for multiplexing may avoid the need to develop new primers and probes.

Evaluation of HCC-QSD with environmental water sample matrices is needed to integrate upstream sample processing. Of the currently available methods for the concentration of viruses in water samples, the glass wool filter method followed by centrifugation or precipitation of viruses appears particularly promising. This method of virus concentration is efficient, inexpensive, and capable of concentrating large volumes of water (e.g. 1500 L drinking water) down to a small volume (approximately 2 ml) suitable for HCC-QSD.

CHAPTER 1
INTRODUCTION

Waterborne disease outbreaks associated with infectious agents continue to occur despite the implementation of advanced water treatment technologies with more than 700 outbreaks documented in the U.S. between 1971 and 2002 (Blackburn, Craun et al. 2004). Further, some experts believe that many more outbreaks occurred that were not reported or recognized. Approximately 8% of these documented waterborne outbreaks were of viral etiology. However, based on the human health outcomes observed, it is speculated that almost half of those outbreaks for which the causative agent was never identified were likely due to enteric viruses (Craun and Calderon 1996). In most cases, viruses were not identified as the cause of the outbreaks due to the limitations of current detection methodology and the failure to examine clinical specimens for viruses.

Over 140 different enteric viruses are known to infect man. The enteric viruses include the enteroviruses (poliovirus, coxsackieviruses, and echoviruses), rotaviruses, hepatitis A and E, Norwalk and Norwalk-like viruses ("noroviruses"), adenoviruses, reoviruses, and others. Compared to bacteria, enteric viruses are more resistant to conventional water and wastewater treatment. Enteroviruses and adenoviruses have frequently been found in both raw source and treated water (Gilgen, Wegmuller et al. 1995; Chapron, Ballester et al. 2000; Greening, Hewitt et al. 2002; Lee and Kim 2002). Reoviruses (Muscillo, Carducci et al. 1997; Spinner and Di Giovanni 2001) and hepatitis A (Schwab, De Leon et al. 1995) have also been found in environmental water samples. Enteric viruses (including poliovirus vaccine strains) have frequently been used as "viral indicators" of fecal contamination (Abbaszadegan, Stewart et al. 1999; Muscillo, La Rosa et al. 2001; Spinner and Di Giovanni 2001). Several of the enteric viruses are good targets for the development of new virus detection methods due to their occurrence in water and their potential risk to public health.

Although adenovirus can affect human populations of all ages, the most severe outcomes are associated with children, immunocompromised, the elderly and those confined in close quarters. Mortality ratios from adenovirus infections have been reported as high as 60% among the immunocompromised (Zahradnik, Spencer et al. 1980; Shields, Hackman et al. 1985). Adenovirus types 40 and 41 are second only to rotavirus as causes of childhood gastroenteritis (Enriquez, Hurst et al. 1995). Besides acute gastroenteritis, adenovirus infections may also result in acute respiratory disease, pneumonia, and eye infections. Adenoviruses are the most common viruses detected in sewage and were detected in 48.3% of the surface water samples collected in the United States during the Information Collection Rule using molecular methodologies (Chapron, Ballester et al. 2000). Adenoviruses are the most resistant waterborne microorganisms to inactivation by ultraviolet light (Thurston-Enriquez, Haas et al. 2003) and survival studies have shown the enteric adenoviruses to be the longest surviving enteric viruses in water (Enriquez, Hurst et al. 1995). There have been several documented waterborne outbreaks, which involved swimming in contaminated recreational waters.

Reoviruses are a common class of enteric viruses capable of infecting a broad range of mammalian species typically with low pathogenicity. Previous studies have shown reoviruses to be common in raw water sources and often found along with other animal viruses (Matsuura, Hasegawa et al. 1984; Spinner and Di Giovanni 2001). This suggests that in addition to the commonly monitored enteroviruses, reoviruses might serve as an informative target for monitoring fecal contamination of drinking water sources (Muscillo, La Rosa et al. 2001; Spinner and Di Giovanni 2001).

Conventional detection of infectious waterborne viruses involves the use of mammalian cell culture, such as the EPA Information Collection Rule (ICR) method. The major strength of cell culture is its ability to detect infectious viruses. However, there are many disadvantages of standard cell culture for the detection of viruses, including cost, two weeks or longer required for completion of the assays, the generic detection of viruses, and the inability to detect some viruses that are not easily cultured. Many different polymerase chain reaction (PCR) methods for the detection of viruses have been developed. PCR methods have the advantages of specific virus detection, including detection of viruses that are not easily cultured (e.g. Norwalk and hepatitis A), speed (same day assays), and reduced cost. Despite these strengths, PCR methods for virus detection do not determine virus infectivity, making the interpretation of data and their health significance unclear. In recent years, integrated cell culture and PCR (ICC-PCR) methods (e.g. (Greening, Hewitt et al. 2002; Reynolds 2004) have attempted to bridge the gap between the two detection strategies. The ICC-PCR approach typically employs a shortened period (e.g. 5 h to 3 days) of cell culture with replicating viruses detected by PCR. While this approach overcomes some of the limitations of cell culture or PCR alone, it is still more time-consuming than desired. Furthermore, most previously developed PCR methods use qualitative detection (i.e. presence or absence) and ICC-PCR methods face challenges for quantitative detection because of virus replication.

In 2003, USEPA, AwwaRF and WERF co-sponsored a "Workshop to Develop a Protocol for Reliable Genetic Methods for the Detection of Viruses for Use in EPA's Water Program" (USEPA 2003). Numerous promising molecular methods were presented, although there was no consensus on the use of a standard set of PCR primers or group of viruses that should be targeted. In addition to the anticipated variations in analytical techniques, several major obstacles to the development and application of molecular methods for the detection of viruses in water were identified. These obstacles included the disparity between PCR results and cell culture infectivity of viruses, the lack of rapid and efficient nucleic acid purification methods, and the lack of quantitative detection. The research performed for this study was aimed at bridging some of these obstacles and providing a foundation for the development of a standard method for the routine monitoring of water for potentially infectious viruses.

RESEARCH OBJECTIVES

The main objective of this research was to further develop a method referred to as host cell capture quantitative sequence detection (HCC-QSD) for the rapid detection of potentially infectious viruses in water. The HCC-QSD method builds on the strengths of cell culture and PCR methods and is compatible with existing water concentration and PCR methods.

The HCC-QSD method uses brief contact with cell culture monolayers to "capture" potentially infectious viruses, a rapid nucleic acid extraction method and real-time quantitative PCR (QSD). The protein capsid that surrounds many animal viruses plays an essential role in viral infection. Upon encountering a potential host cell, the viral capsid must interact with one or more cell surface receptors. In addition, for successful infection and replication, viruses must have intact nucleic acids. Therefore, attachment to host cells and intact nucleic acid are two requirements for viral infection and form the basis of the HCC-QSD strategy. For the host cell capture (HCC) step, cell culture monolayers were incubated with virus samples for three hours or less to "capture" viruses, followed by washing to remove non-adsorbed viruses and sample debris. RNA from "captured" viruses was then extracted using a commercial RNA purification kit and quantified using QSD. These conditions, combined with the specificity of molecular detection, were used to selectively recover, detect, and quantify potentially infectious viruses. In contrast to ICC-PCR methods, HCC-QSD does not require viral replication in cell culture.

QSD requires the use of primers similar to those used in conventional PCR; however, unlike conventional PCR, QSD also requires an oligonucleotide probe labeled with 5′ reporter and 3′ quencher fluorescent dyes and a thermal cycler equipped with a fluorometer. During each cycle of amplification, if the target of interest is present, the probe specifically anneals to the target amplicon between the forward and reverse primer sites. Due to the 5′ nuclease activity of *Taq* DNA polymerase, the probe is cleaved during the polymerization step (amplicon formation) resulting in an increase in reporter fluorescence detected by the instrument. Target signal increases in direct proportion to the concentration of the target amplicon being formed. The threshold cycle (C_T) is the fractional PCR cycle number at which a significant increase in target signal fluorescence above baseline is first detected for a sample. By using amplification standards consisting of known quantities of target nucleic acid or organisms to generate a standard curve, the starting copy number of nucleic acid targets or target organisms for each sample can be estimated.

CHAPTER 2
EVALUATION OF DIFFERENT HOST CELL LINES FOR THE CAPTURE OF ENTERIC VIRUSES AND OPTIMIZATION OF CAPTURE CONDITIONS

INTRODUCTION

In previous work conducted by the Project Team as part of the Water Environment Research Foundation Project 00-HHE-2b (Di Giovanni, Karim et al. 2002), only the Buffalo Green monkey kidney (BGMK) cell line was evaluated for HCC-QSD. The BGMK cell line is routinely used for the culture of enteroviruses (including coxsackievirus, poliovirus, and other enteroviruses) and is the cell line used in the USEPA ICR method. Vaccine strains of poliovirus are easily cultured with this cell line and can be enumerated using a plaque-forming assay. We evaluated the efficiency of HCC using a model system of BGMK cell monolayers in microplates and vaccine strain poliovirus. Quantitation of capture efficiency was determined by plaque assay. Three different contact ("capture") times were evaluated: two hours, 15 minutes and no contact (less than 30 seconds). In these trials, approximately 25 PFU of poliovirus in 100 µl of PBS was inoculated onto replicate cell monolayers. Inoculated monolayers were incubated at 37 °C in a humidified CO_2 incubator and rocked every 15 minutes to distribute the inocula. To determine the numbers of viruses captured by the monolayers, the inocula were aspirated from one set of replicate wells for each treatment group, and the monolayers washed with 3 ml of PBS to remove unadsorbed virus. The "captured" viruses remaining on the washed monolayers were enumerated by plaque assay and expressed as the percentage of the unwashed controls. The two-hour capture time was most efficient, with a mean of 93% of the viruses captured by the monolayers. Not surprisingly, reduced capture was observed when the capture time was lessened, with a mean of 71% of the viruses captured after 15 minutes. No capture time (less than 30 seconds) resulted in only 4% capture efficiency of the viruses by the monolayers. Therefore, plaque assay results indicated efficient capture by the host cell monolayers with capture times of 15 minutes to 2 hours.

In the present study, similar experiments were performed to determine HCC-QSD method recoveries. Cell lines for evaluation included Buffalo Green monkey kidney (BGMK) cells, Vero African Green monkey kidney cells, Graham 293 human kidney cells, and FRhK-4 Rhesus monkey kidney cells. The Graham cell line was difficult to start but eventually cultured, although growth stopped after several passages. The FRhK-4 cells received from ATCC were not able to be cultured, despite following ATCC's recommendations and were not able to be evaluated. Therefore, we used the BGMK and Vero cell lines for experiments performed in this study.

Candidate viruses included poliovirus type 3 (attenuated), hepatitis A, adenovirus serotypes 40 and 41, and mammalian reovirus type 3 (Dearing strain). Because the appropriate cell lines for adenovirus serotypes 40 and 41 and hepatitis A were not cultured, we used poliovirus type 3 (attenuated) and mammalian reovirus type 3 (Dearing strain). Various HCC monolayer formats were evaluated to optimize the capture efficiency. These formats included the use of 12-well microplates, different ratios of sample volume to cell monolayer surface area, and different inoculation media. The effect of different contact times (e.g. 0, 30, 60 and 120 minutes) on efficiency of HCC was investigated. For the majority of experiments, the model system of the BGMK cell line and poliovirus was used. Table 2.1 describes the viruses, cell lines, and PCR primers and probes that were used in this study.

After optimizing capture conditions, trials were performed to evaluate the ability of the BGMK and Vero cell lines to capture (adsorb) poliovirus and reovirus. For these trials, a minimum of five replicates for treatment groups were performed. Positive extraction controls for which the inocula were not washed off of monolayers and uninoculated negative controls were also included.

Table 2.1

Viruses, cell lines, and PCR primers and probes used in this study

Virus	Cell line	Primer and TaqMan Probe Sequences	Reference
poliovirus 3 (Fox), attenuated (ATCC VR-193)	Buffalo Green monkey kidney (BGMK, USEPA)	EV1 5-GATTGTCACCATAAGCAGC-3 EV2 5-CCCCTGAATGCGGCTAATC-3 5-FAM-CGGAACCGACTACTTTGGGTGTCCGT-TAMRA	(Monpoeho, Coste-Burel et al. 2002)
mammalian reovirus 3 (Dearing) (ATCC VR-824)	Vero African Green monkey kidney (ATCC CCL-81)	Reo3F 5-TGATTTCCATTACTTCTGCTGCTT-3 Reo3R 5-TCCTGTTCACGATTCCATCAGAT-3 5-FAM-CGCCGTGCTTGACCACCCACTT-TAMRA-3	(Uchiyama and Besselsen 2003)

HOST CELL CAPTURE QUANTITATIVE SEQUENCE DETECTION (HCC-QSD)

Methods

Cell Monolayers

For BGMK cells, Minimal Essential Medium (MEM) containing 10% fetal bovine serum, sodium bicarbonate, and antibiotics was used for maintenance medium. For Vero cells, MEM with 10% fetal bovine serum and sodium bicarbonate was used for maintenance medium. Cell monolayers were cultured in 12-well cell culture plates with each well seeded with 2.0 ml of 5.25×10^4 cells ml^{-1} in maintenance medium. Plates were incubated at 36°C in a 5% CO_2 atmosphere. Virus spiking and RNA extraction were performed on three to five day-old confluent monolayers.

Host Cell Capture and RNA Extraction

Five or six replicates of no contact, 30, 60, 120, and 180 minute capture times were used to evaluate the effect of capture times on HCC. Monolayers were washed with 2.0 ml of PBS immediately prior to spiking to remove cell culture medium. Poliovirus spikes were added in a total of 100 μL PBS and monolayers were incubated at 36°C in 5% CO_2 with rocking every 15 minutes. For positive and negative controls, 100 μL of PBS was added to prevent drying of the monolayers. After the desired capture time, the spike suspensions were aspirated, the monolayers washed with 2.0 ml of PBS to remove unadsorbed viruses, and the PBS aspirated. Positive extraction controls were then prepared by inoculating monolayers with an equivalent amount of virus in a total of 10 μL PBS (reduced volume for compatibility with the RNA extraction kit) and RNA extracted without aspirating or washing off the spike. Qiagen QIAshredder columns and RNeasy kit were used for extraction of total RNA from monolayers. RNA was eluted from the purification columns using 50 μL of RNase-free molecular grade water and samples were stored immediately at -70°C until assayed by QSD. As the experiments became larger and more complex, extractions were performed using the Qiagen RNeasy 96 plates with a vacuum manifold for increased throughput.

As mentioned previously, culture of all proposed cell lines was not accomplished. Therefore HCC-QSD work used Buffalo Green monkey kidney (BGMK, USEPA) and Vero African Green monkey kidney (ATCC CCL-81) cell lines. Propagation of attenuated poliovirus 3 (Fox strain, ATCC VR-193) and mammalian reovirus 3 (Dearing strain, ATCC VR-824) were accomplished with the cultured cell lines and used as models. Two HCC optimization experiments for each virus and cell line system were performed. The poliovirus and BGMK cell line model system experiment consisted of five replicates for each treatment, positive extraction control and method blanks to serve as negative controls. Seeded monolayers were spiked with 100 PFU of poliovirus.

For reovirus and the Vero cell line, six replicates for each contact time as well as positive extraction controls and method blanks were used. Seeded monolayers were spiked with 1000 PFU of reovirus. Again, two individual experiments were run to demonstrate reproducibility of significant differences between contact times rather than comparing C_T values directly. The host cell capture capability of BGMK cells for reovirus was also evaluated. This experiment was composed of six replicates for all capture times, positive extraction controls and method blanks. Seeded monolayers were inoculated with 1000 PFU spikes.

Quantitative Sequence Detection

QSD was performed to evaluate and optimize HCC conditions. For the poliovirus and BGMK cell model system, pan-enterovirus primers and TaqMan probe were used (Monpoeho, Coste-Burel et al. 2002). Enterovirus QSD positive controls were purified poliovirus 3 RNA, and no template negative controls were included in each QSD run. Enterovirus QSD was performed using an Applied Biosystems 7000 Sequence detector with amplification conditions as follows: reverse transcription for 45 min at 42°C, initial denaturation for 10 min at 95°C, followed by 45 cycles of denaturation at 95°C for 30 sec and annealing at 60°C for 1 min. (Monpoeho, Dehee et al. 2000; Monpoeho, Coste-Burel et al. 2002).

QSD for mammalian reovirus was adapted from a two-tube reverse transcription quantitative PCR protocol. The protocol was modified to a one-tube reverse transcription QSD that resulted in a 72 bp product. Reovirus QSD positive controls were composed of purified reovirus RNA, and no template negative controls were included in each QSD run. QSD for reovirus was also performed using an Applied Biosystems 7000 Sequence detector Amplification with the following conditions: reverse transcription for 45 min at 42°C, initial denaturation for 10 min at 95°C, followed by 45 cycles of denaturation at 95°C for 30 sec, and annealing at 60 °C for 1 min (Uchiyama and Besselsen 2003).

Data Analysis and Statistics

Threshold cycle (C_T) data for experiments were expressed as mean and standard deviation of replicates. To determine the statistical significance of results, data were analyzed by one-way analysis of variance (ANOVA) and post hoc Bonferroni pairwise mean comparisons using STATA version 9 (College Station, Texas). An alpha level of 0.05 was used for all statistical tests. The threshold cycle (C_T) is the fractional PCR cycle number at which a significant increase in target signal fluorescence above baseline is first detected for a sample. The lower the C_T value, the earlier amplification has occurred.

Results

Evaluation of Different Capture Times for the Optimization of Host Cell Capture

For poliovirus 3, increased capture time typically resulted in increased capture of viruses, as seen by lower QSD threshold cycle (C_T) values. Amplification curves for replicates consistently appeared normal and robust, negative controls (method blanks and PCR no-template controls) were negative even after 45 cycles of amplification. Results from the evaluation of different capture times on the HCC of poliovirus 3 using the BGMK cell line are provided in Table 2.2. Results are presented as the change in C_T values (Δ C_T) compared to the positive extraction controls, with the smaller the Δ C_T value, the greater capture of virus. Differences in the range of C_T values between the two experiments are most likely related to the use of two different virus stocks and differences in QSD threshold settings for each run.

Table 2.2
Evaluation of different capture times on the HCC of poliovirus 3 using the BGMK cell line

Capture Time	Mean Δ $C_T \pm$ SD from positive extraction controls (n = 5)	
	Experiment 1	Experiment 2
Positive extraction controls	0.00 ± 0.35	0.00 ±0.36
No contact (<1 minute)	NA*	5.27 ± 1.24
30 minutes	3.27 ± 0.65	2.69 ± 0.41
1 hour	2.45 ± 0.86	2.45 ± 0.66
3 hours	1.69 ± 0.46	1.41 ± 0.77

* No Amplification

One-way analysis of variance (ANOVA) of C_T values revealed that a capture time as short as 30 minutes recovered significantly more viruses ($P < 0.05$) than no capture time (less than 1 minute). Similarly, a 3-hour capture time resulted in significantly more viruses captured than a 30 minute capture time in both experiments ($P = 0.002$ and $P = 0.01$, respectively). While mean C_T values were lower for 3-hour capture samples compared to 1-hour capture samples, the differences were not very different ($P = 0.09$ and $P = 0.05$, respectively). It could be speculated that even longer capture times would result in greater virus capture. By comparison of the C_T values for RNA extraction controls (which provide the lowest possible C_T values), 30 minute, and 3-hour capture samples, it appears that maximum capture would occur after approximately 6 hours. However, in light of the goal to develop a rapid, single-day assay, two to three hours were the longest practical capture times used for the remainder of the study.

For reovirus experiments, amplification curves for samples appeared consistently normal and robust and negative controls (method blanks and PCR no-template controls) were negative even after 45 cycles of amplification. Tables 2.3 and 2.4 present the change in mean threshold cycle C_T value when compared to positive extraction controls for reovirus HCC experiments using the Vero and BGMK cell lines, respectively. In contrast to the poliovirus and BGMK experiments, in most cases increasing the capture times did not result in increased capture of reovirus using either the Vero or BGMK cell lines (with the exception of Expt. 2 with Vero cells).

Table 2.3

Evaluation of different capture times on the HCC of reovirus using the Vero cell line

Capture Time	Mean $\Delta C_T \pm$ SD from positive extraction controls (n = 6)	
	Experiment 1	Experiment 2
Positive extraction controls	0.00 ± 0.36	0.00 ± 0.01
No contact (<1 minute)	2.61 ± 0.87	4.03 ± 2.46
30 minutes	5.08 ± 2.89	2.03 ± 0.35
1 hour	5.90 ± 3.23	4.47 ± 0.97
2 hours	5.11 ± 3.64	-1.14 ± 0.33

Table 2.4
Evaluation of different capture times on the HCC of reovirus using the BGMK cell line

Capture Time	Mean Δ C_T ± SD from positive extraction controls (n = 6)
Positive extraction controls	0.00 ± 0.79
No contact (<1 minute)	3.36 ± 1.00
30 minutes	9.53 ± 3.04
1 hour	3.87 ± 3.23
2 hours	4.58 ± 2.19

Conclusions and Recommendations

For the BGMK cells and poliovirus model, a trend of increased virus capture with increased capture time was clearly demonstrated as hypothesized, with the maximum 3-hour capture time resulting in the highest capture of virus. Duplicate experiments were performed to evaluate reproducibility of capture results, and the same trend of increased capture with increased capture time was observed for both experiments.

In contrast, while reovirus were captured by cell monolayers, reovirus experiments did not show the same trend of increasing capture with increasing time for either the Vero or BGMK cell lines. However, for experiments using reovirus and the BGMK cell line in particular, sloughing of monolayers was observed which may have introduced variability in the results. Monolayers should maintain intact through HCC, therefore further experiments to evaluate the use of cell culture media instead of PBS for inoculation and washing of monolayers were performed as described in the following section.

VOLUME AND INOCULATION MEDIUM OPTMIZATION EXPERIMENTS

Methods

A comparison of different inoculation volumes and inoculum preparation in serum-free MEM versus PBS was performed in response to the observed loss of cell monolayers during the capture step. Some of the monolayers appeared to be drying out during the capture step incubation or sloughing off when washed after incubation. BGMK monolayers were inoculated with PBS and MEM basal media in order to determine if the PBS was responsible for the damaged monolayers and the sloughing of cells.

In addition, to assess the effect of inoculum volume on monolayer integrity, an experiment using poliovirus and BGMK cells was performed using the same virus concentration in 100 μl and 150 μl inoculum volumes. The experiment was performed with all capture times in order to observe any time-dependent differences. This was done on a smaller scale with three replicates for each volume and contact time in order to incorporate all capture times in the same experiment.

Results

BGMK monolayers remain intact when inoculated with serum-free MEM medium as opposed to inoculation with PBS (Figures 2.1 and 2.2). These figures provide visual evidence to demonstrate that inoculum preparation in serum-free MEM medium prevents the rounding and sloughing of cells compared to PBS.

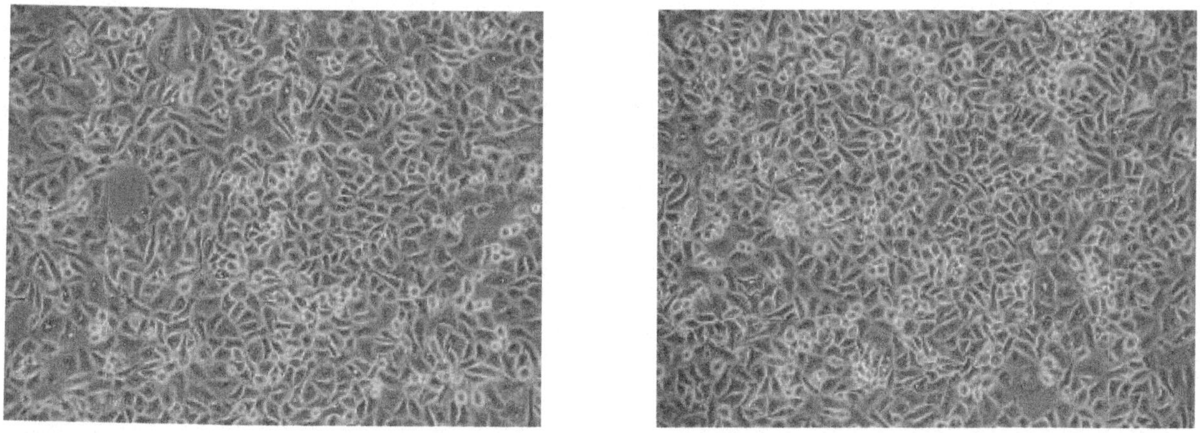

Figure 2.1 12-well BGMK monolayers after 2-hour capture time with 100 μl of serum-free MEM medium

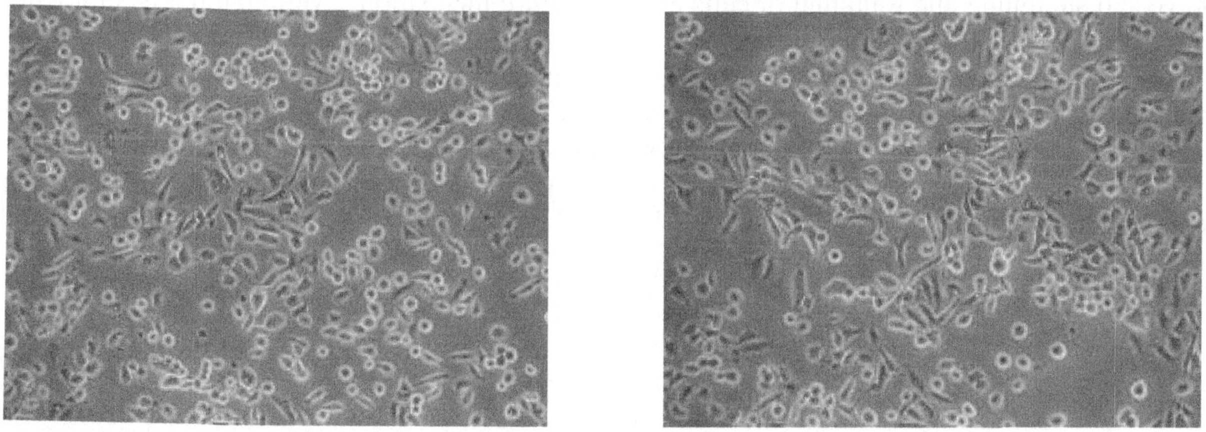

Figure 2.2 12-well BGMK monolayers after 2-hour capture time with 100 μl of PBS

Table 2.5 presents the mean and standard deviations for each capture time using the two different inoculum volumes. One-way analysis of variance (ANOVA) of C_T values revealed no significant difference ($P > 0.05$) in the efficiency of virus capture between the use of 100 μl or 150 μl inoculum volumes for any of the capture times.

Table 2.5
Volume comparison experiment using different inoculum volumes of poliovirus 3 on BGMK cell line

Capture Time	Mean ± SD C_T value (n = 3)	
	100 µl	150 µl
Positive extraction controls	26.36 ± 0.32	27.34 ± 2.04
No contact (<1 minute)	35.35 ± 4.60	31.86 ± 2.51
30 minutes	29.77 ± 0.41	29.46 ± 0.16
1 hour	30.04 ± 0.22	30.36 ± 0.15
2 hours	29.26 ± 0.13	29.02 ± 0.26

Conclusions and Recommendations

The original protocol for HCC-QSD included washing monolayers with PBS prior to inoculation and extraction, and preparation of inocula in PBS. PBS washing was used to remove all traces of cell culture maintenance media which could potentially inhibit PCR. Results demonstrated that preparing inocula in serum-free MEM medium, particularly for BGMK cells, prevented sloughing and rounding of cells during capture incubation. Although not as dramatic, a similar beneficial effect of MEM versus PBS was also observed for Vero cells (data not shown). For volume optimization experiments, the lower 100 µl inoculation volume was expected to result in partial loss of monolayers due to drying. However, visual evidence and HCC-QSD data results indicated that there was no difference between the 100 µl and 150 µl volumes. Consequently, 100 µl inocula in serum-free MEM medium were used for the remainder of the experiments in this project.

CHAPTER 3
HOST CELL CAPTURE QUANTITATIVE SEQUENCE DETECTION (HCC-QSD) OF VIRUSES AND EVALUATION OF STANDARD AND ADVANCED NUCLEIC ACID PURIFICATION METHODS

Introduction

Real-Time Quantitative PCR (QSD) Quantitation of Viruses

Our preliminary trials of the HCC-QSD method used pan-enterovirus primers and a dual-labeled fluorogenic (TaqMan) probe. While most PCR-based methods for the generic detection of enteroviruses in water rely upon these same or similar primers, many PCR protocols for the specific detection of different viruses exist. Published primer sets and TaqMan probes were evaluated for their thermal characteristics (e.g. melting temperatures), amplicon length, and sequence to determine their suitability for HCC-QSD. Selected primers and probes were evaluated using purified viral nucleic acid standards and standard curves were constructed for the use in quantifying viruses in HCC samples.

Nucleic Acid Purification

As anticipated, QSD enumeration of virus HCC samples (containing BGMK cell RNA and DNA) and purified viral RNA differed. This was likely due to interference by RNA (and possibly DNA) extracted from the BGMK cell culture monolayers. To overcome this adverse effect, we proposed the use of a commercially available immunomagnetic separation system and oligonucleotide probe capture of specific pan-enterovirus RNA or adenovirus DNA. Oligonucleotide probes (PCR primers, selected after preliminary evaluation as described above) were synthesized and labeled with biotin at the 5′ end. These biotinylated probes/primers were mixed with nucleic acids extracted from host cell monolayers obtained using commercial extraction kits (Qiagen Viral RNA and DNA kits) and incubated under oligo hybridization conditions. Probe captured viral nucleic acids were selectively recovered from the host cell/virus nucleic acid mixture by the addition of commercially available streptavidin-labeled immunomagnetic beads (Dynal Biotech). The viral nucleic acid-immunomagnetic bead complexes were recovered from the samples using a magnetic particle concentrator and unbound primer and interfering host cell nucleic acid removed by washing.

This technology has been used for numerous years by the biotechnology industry and clinical/food microbiology field but is just beginning to be evaluated for environmental microbiology applications. Advantages of this approach include flexibility of the purification platform for specific viruses (and potentially mixtures of different viruses using the same protocol) and additional removal of potential PCR inhibitors. The previously developed standard HCC-QSD nucleic acid extraction procedure (Di Giovanni, Karim et al. 2002) was compared to the probe capture immunomagnetic separation procedure using QSD. Each nucleic acid extraction and purification method was evaluated using a minimum of five replicates.

EVALUATION OF RNA EXTRACTION KITS

Methods

Two commonly used RNA extraction kits, the Qiagen RNeasy and Qiagen Viral RNA kits, were evaluated in order to determine which produced the highest quality RNA extract from virus seeded cell monolayers.

BGMK Monolayers

Minimal Essential Medium (MEM) containing 10% fetal bovine serum, sodium bicarbonate and antibiotics was used for maintenance medium as previously described. Cell monolayers were cultured in 12-well cell culture plates with each well seeded with 2.0 ml of 5.25 X 10^4 cells ml^{-1} in maintenance medium. Plates were incubated at 36°C in a 5% CO_2 atmosphere. Virus spiking and RNA extractions were performed on three to five day-old confluent monolayers.

Virus Seeding and RNA Extraction

Monolayers were washed with 2.0 ml of PBS immediately prior to RNA extraction. Approximately 150 PFU of poliovirus was added to seed monolayers. This was achieved by adding 10 µL of a titered virus stock (in PBS) plus 130 µL PBS to monolayers for the Viral RNA kit. Only 10 µL of virus stock plus 90 µL PBS was added to monolayers for the RNeasy kit. The Qiagen RNeasy and Viral RNA kits were used for extraction following manufacturer protocols. For the RNeasy kit this included homogenization of cell lysates using a syringe or QIAshredder column. RNA was eluted from the purification columns using 50 µL of Qiagen AVE buffer for the Viral RNA kit or 50 µL of RNase-free molecular grade water for the RNeasy kit. RNA extracts were immediately stored at -70°C until assayed by QSD.

Quantitative Sequence Detection

The pan-enterovirus primers and TaqMan probe were used (Monpoeho, Coste-Burel et al. 2002) for QSD to assess the quality of the RNA extracts. We previously used these primers and probe in a Water Environment Research Foundation (WERF) project aimed at the initial development of the HCC-QSD assay (Di Giovanni, Karim et al. 2002). As a control, QSD reaction cocktails were seeded with RNA extracted from unseeded monolayers and purified poliovirus RNA extracted from approximately 150 PFU.

Results

The threshold cycle (C_T) is the fractional PCR cycle number at which a significant increase in target signal fluorescence above baseline is first detected for a sample. The lower the C_T value, the earlier amplification has occurred. Comparison of C_T values for samples seeded with the same amount of target (i.e. virus or nucleic acid) was used to assess purification efficiency or sample quality. In each experiment, triplicate no-template negative controls were performed and all tested negative. RNA extracted from unseeded BGMK monolayers using both extraction kits also tested negative. Amplification plots and C_T values of samples from each experiment are presented in Figures 3.1, 3.2, 3.3 and 3.4 and Tables 3.1, 3.2 and 3.3.

Experiment 1

Five replicates seeded with 150 PFU poliovirus 3 and unseeded monolayers were extracted using the RNeasy and Viral RNA kits and extracts analyzed by QSD. C_T values of samples are presented in Table 3.1. The QSD amplification plot of samples is presented in Figure 3.2.

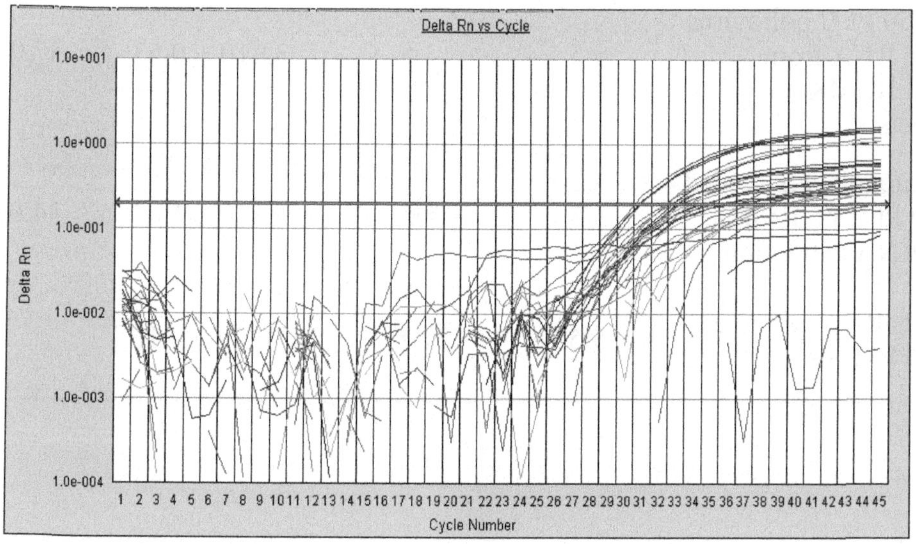

Figure 3.1 QSD amplification plot of experiment 1 samples for the evaluation of Qiagen RNeasy and Viral RNA extraction kits

Table 3.1
Mean and standard deviation threshold cycle (C_T) values for experiment 1

Sample type	Number of positive QSD replicates/total number replicates		Mean ± SD C_T Value	
	Qiagen Viral RNA kit	Qiagen RNeasy kit	Qiagen Viral RNA kit	Qiagen RNeasy kit
QSD reaction cocktails seeded 150 PFU poliovirus RNA and RNA from unseeded BGMK monolayers	5/5	3/5	33.10 ± 0.53	37.07 ± 1.60
BGMK monolayers seeded with 150 PFU poliovirus prior to RNA extraction	3/5	5/5	37.10 ± 1.63	34.07 ± 0.89
Poliovirus RNA extracted from 150 PFU only	5/5	ND*	30.79 ± 0.16	ND*

*Not Done

Experiment 2

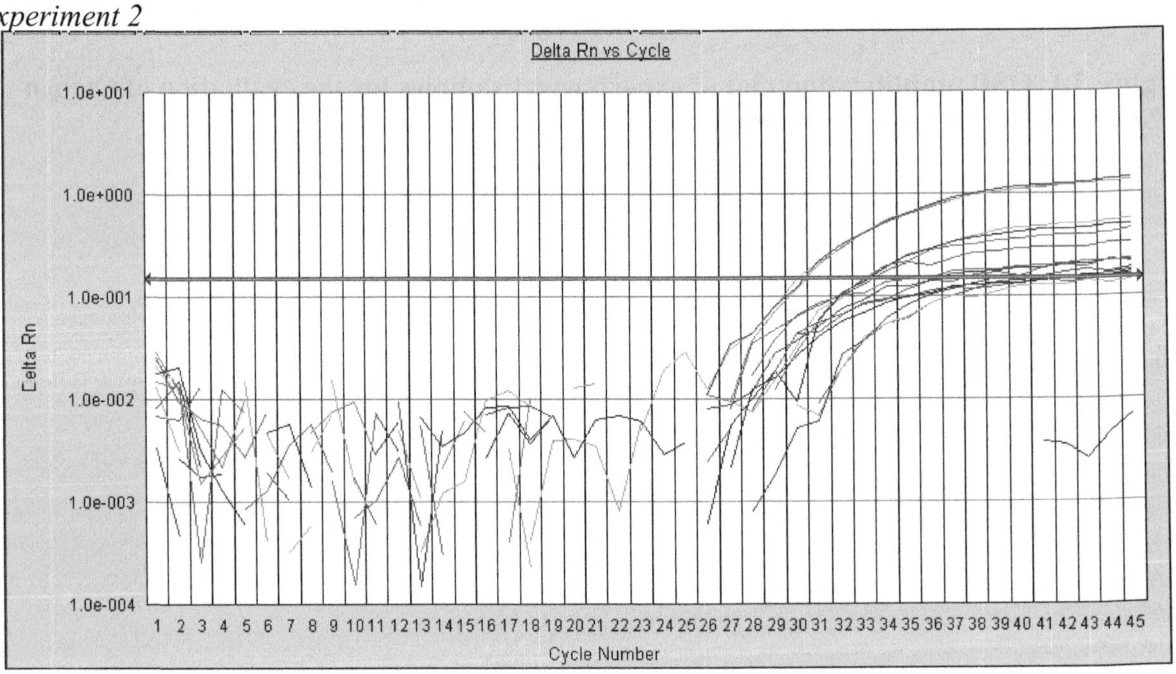

Figure 3.2 QSD amplification plot of experiment 2 samples for the evaluation of Qiagen RNeasy and Viral RNA extraction kits

Table 3.2
Mean and standard deviation threshold cycle (C_T) values for experiment 2

Sample type	Number of positive QSD replicates/total number replicates		Mean ± SD C_T Value	
	Qiagen Viral RNA kit	Qiagen RNeasy kit	Qiagen Viral RNA kit	Qiagen RNeasy kit
QSD reaction cocktails seeded 150 PFU poliovirus RNA and RNA from unseeded BGMK monolayers	3/3	3/3	33.19 ± 0.06	39.24 ± 3.97
BGMK monolayers seeded with 150 PFU poliovirus prior to RNA extraction	2/3	3/3	41.63 ± 1.23	37.18 ± 4.35
Poliovirus RNA extracted from 150 PFU only	3/3	ND*	30.28 ± 0.16	ND*

*ND, not done.

Experiment 3

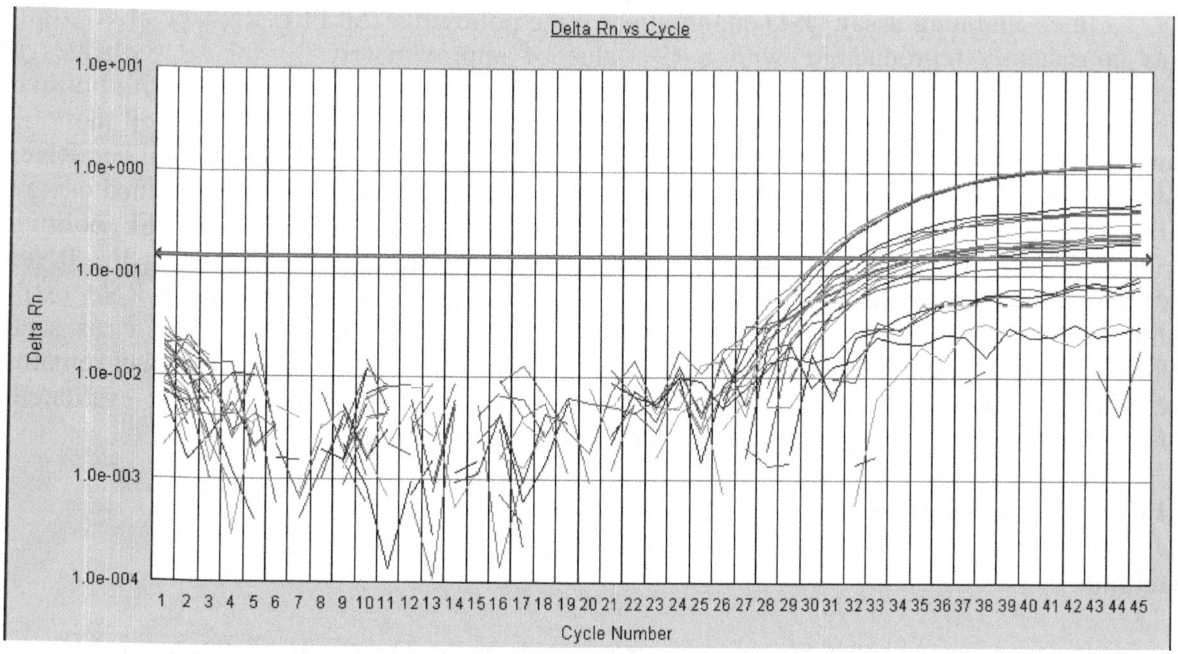

Figure 3.3 QSD amplification plot of experiment 3 samples for the evaluation of Qiagen RNeasy and Viral RNA extraction kits

Table 3.3
Mean and standard deviation threshold cycle (C_T) values for experiment 3

Sample type	Number of positive QSD replicates/total number replicates		Mean ± SD C_T Value	
	Qiagen Viral RNA kit	Qiagen RNeasy kit	Qiagen Viral RNA kit	Qiagen RNeasy kit
QSD reaction cocktails seeded 150 PFU poliovirus RNA and RNA from unseeded BGMK monolayers	4/5	5/5	32.47 ± 0.43	35.48 ± 1.59
BGMK monolayers seeded with 150 PFU poliovirus prior to RNA extraction	0/5	4/5	NA[1]	37.35 ± 4.78
Poliovirus RNA extracted from 150 PFU only	5/5	ND[2]	30.53 ± 0.16	ND

[1]NA, not applicable. [2]ND, not done.

Conclusions and Recommendations

Inter- and intra-assay QSD quantitation of the poliovirus 150 PFU RNA positive controls was consistently reproducible, with a C_T value of approximately 30 for all replicates and experiments. As observed in the WERF project, QSD reaction cocktails seeded with poliovirus RNA and RNA from unseeded BGMK monolayers revealed an adverse effect of cell monolayer nucleic acids on quantitation of viral RNA. This was true for both the Viral RNA kit and RNeasy kits. However, for this type of sample, the lowest C_T values were consistently obtained using the Viral RNA extraction kit. Conversely, for BGMK monolayers seeded with 150 PFU poliovirus prior to RNA extraction, the lowest C_T values were consistently obtained using the RNeasy extraction kit. Thus, it appears that the Viral RNA kit provides the highest quality RNA extract, while the RNeasy kit provides the most reliable extraction of viral RNA from virus seeded BGMK monolayers. Therefore, the RNeasy kit was the likely choice for the remaining experiments. However, the use of carrier RNA in the Viral RNA kit was first evaluated to ensure that this was not responsible for differences in kit performance.

EFFECT OF CARRIER RNA

Methods

It was suspected that the use of carrier RNA in the Qiagen Viral RNA kit AVL lysis buffer combined with the RNA extracted from the cell monolayer may have been overloading the purification columns. This could explain the poor amplification of poliovirus seeded BGMK samples observed when using the Viral RNA kit. To investigate this, AVL buffer was prepared with 1X, 0.5X and no carrier RNA and extraction efficiency and extract quality determined using QSD as described in the prior experiments.

Results

Statistical analysis using one-way analysis of variance (ANOVA) indicated no significant difference between any of the carrier RNA concentrations used. Even for the highest and lowest levels of carrier RNA, i.e. 100% versus none, there was no significant difference.

Conclusions and Recommendations

No significant difference in QSD amplification was observed between any of the concentrations of carrier RNA. Therefore, due to the lower performance of the Viral RNA kit compared to the Qiagen RNeasy kit, the RNeasy kit was used for the remaining HCC-QSD experiments.

EVALUATION OF QIASHREDDER AND MERCAPTOETHANOL

Methods

Qiagen QIAshredders are spin columns that are used for the initial homogenization of cell lysates to ensure nucleic acids are available for adsorption to purification columns in subsequent steps of extraction. The manufacturer recommends using QIAshredders or some other homogenization technique for cell lysate samples prior to loading samples onto RNeasy spin columns. An experiment was performed to evaluate the benefit of using the QIAshredder columns. The HCC-QSD protocol was followed for extraction of RNA from poliovirus 3 seeded BGMK cell monolayers.

The extraction kit instructions also include the addition of β-mercaptoethanol to the kit lysis buffer. Sample processing with or without β-mercaptoethanol addition to the Qiagen RNeasy RLT lysis buffer was also incorporated into the experiment to test whether it affected RNA extraction efficiency from seeded monolayers.

Results

One-way analysis of variance (ANOVA) of the C_T values resulted in a significant *P*-value of 0.047 for the use of QIAshredders and the use or exclusion of β-mercaptoethanol in the lysis buffer. However, Bonferroni post-hoc group comparison revealed no statistical difference for the use of QIAshredders and only a borderline significance when using β-mercaptoethanol in lysis buffer ($P = 0.054$). There was no significant difference ($P > 0.05$) between C_T values of samples extracted with or without β-mercaptoethanol and without the use of QIAshredders. The mean C_T values for all treatment groups were very close, with less than 1 C_T value difference (Table 3.4).

Table 3.4
HCC-QSD comparison of RNA extraction variations

RNeasy kit variations	Mean ± SD C_T Value
QIAshredder + RLT Lysis Buffer	28.03 ± 0.24
QIAshredder + RLT Lysis Buffer with β-mercaptoethanol	28.36 ± 0.49
RLT Lysis Buffer	27.79 ± 0.44
RLT Lysis Buffer with β-mercaptoethanol	27.58 ± 0.11

Conclusions and Recommendations

The use of QIAshredders and β-mercaptoethanol resulted in no significant difference in RNA yield. This was likely due to the readily lysed cell monolayer format of the HCC-QSD assay as opposed to more difficult to lyse tissue samples that can be used with the kit (e.g. biopsy samples or mouse tails). Due to its toxicity and potential risk to laboratory personnel, β-mercaptoethanol is not recommended for inclusion in the RLT lysis buffer. Therefore, the RNeasy kit was used without QIAshredders and β-mercaptoethanol.

OLIGO CAPTURE PURIFICATION

Introduction

Four oligo capture immunomagnetic separation (IMS) experiments were performed to evaluate this strategy for the selective separation of viral RNA from BGMK monolayer RNA. Several different conditions were evaluated, including a standard one-tube reverse transcription QSD, a separate reverse transcription step, effects of Microcon concentration of cDNA, and the effects of BGMK monolayer RNA.

Methods

BGMK Monolayers

BGMK monolayers were cultured in 12-well cell culture plates. Each well was seeded with 2.0 ml of 5.25 X 10^4 cells ml^{-1} in MEM maintenance medium containing 10% fetal bovine serum. Plates were incubated at 36°C and 5% CO_2 atmosphere. Virus spiking and RNA extraction was performed on three to five day-old monolayers.

Host Cell Capture and RNA Extraction

Monolayers were washed with 2.0 ml of PBS immediately prior to spiking to remove cell culture medium. Poliovirus spikes were added in a total of 100 µL PBS and monolayers were incubated at 37°C in 5% CO_2 with rocking every 15 minutes. For positive and negative controls, 100 µL of PBS was added to prevent drying of the monolayers. After the desired capture time, the spike suspensions were aspirated, the monolayers washed with 2.0 ml of PBS to remove unadsorbed viruses, and the PBS aspirated. Host Cell Capture RNA positive extraction controls consisted of an equivalent PFU of poliovirus in a total of 10 µL PBS, reduced volume for compatibility with the RNA extraction kit, spiked onto monolayers and RNA extracted without aspirating or washing off the spike. Negative controls consisted of RNA extracted from unspiked monolayers. Qiagen QIAshredder columns and the RNeasy kit were used for extraction of total RNA from monolayers. RNA was eluted from the purification columns using 50 µL of RNase-free molecular grade water and samples were stored immediately at -70 °C until assayed by QSD.

Advanced Nucleic Acid Purification Using Biotin Oligo Capture and Immunomagnetic Separation (IMS)

As opposed to our typical single-tube RT-PCR reaction, a separate reverse transcription step was performed for these experiments. Biotin-labeled pan-enterovirus PCR primers were used to specifically label poliovirus cDNA. Reverse transcription cocktails containing biotin-labeled pan-enterovirus PCR primers were spiked with purified poliovirus RNA (extracted from 100, 500 and 1000 PFU) and with or without BGMK cell monolayer RNA. RT was performed for 45 minutes at 42°C, followed by inactivation at 95°C for 10 minutes. Free (unincorporated) biotin-labeled oligos were removed from the samples using Microcon YM-100 100,000 molecular weight cutoff filters. Biotin-labeled viral cDNA was captured using the Dynal Dynabeads kilobaseBINDER Kit with streptavidin M-280 Dynabeads and a 2 hour room temperature capture time.

Since the biotin-streptavidin bond is very stable and is not broken even by 95°C heating, the beads were added directly to the QSD reaction mix rather than trying to remove the captured cDNA from the IMS beads. The PCR is able to proceed by priming the complementary DNA strands directly off of the cDNA bound to the IMS beads.

Quantitative Sequence Detection

The pan-enterovirus primers and TaqMan probe (Monpoeho, Coste-Burel et al. 2002) described elsewhere in this report were used for our poliovirus and BGMK cell model system. QSD positive controls (purified poliovirus 3 RNA) and no template negative controls were included in each QSD run.

Results

Four oligo capture immunomagnetic separation (IMS) experiments were performed to further develop this approach for the selective separation of viral RNA from BGMK monolayer RNA. The first experiment compared our standard one-tube reverse transcription QSD method to a separate reverse transcription followed by Microcon concentration and IMS purification of the biotin-labeled cDNA. Purified poliovirus RNA (from 1000 PFU) and molecular grade water (no BGMK RNA) was used for this experiment. The one-tube QSD samples had significantly better amplification ($P < 0.05$, $n = 5$), with a mean and standard deviation C_T of 25.51 ± 0.18 compared to 29.15 ± 0.28 for the separate tube reverse transcription, IMS-purified samples. One possible explanation for the difference could be the loss of cDNA during the Microcon concentration step. Therefore, a subsequent experiment was performed to investigate this possibility.

To evaluate the effect of the Microcon concentration, purified poliovirus RNA (from 500 PFU) and molecular grade water (no BGMK RNA) was used for a separate tube reverse transcription step. The resulting cDNA was concentrated using the Microcon filters, but the entire concentrate was added to a PCR reaction mix, rather than attempting further purification using IMS. These samples were compared to the standard one-tube QSD seeded with the same amount of viral RNA. Results indicated that the separate RT and Microcon concentration of samples had a negligible effect on amplification compared to the standard one-tube QSD ($P = 0.11$, $n = 5$).

Since it appeared that the IMS recovery of the biotin-labeled cDNA was the critical step in this purification approach, we performed another repeat experiment (Experiment 3) comparing the separate reverse transcription and oligo capture IMS purification to the standard QSD method. For this experiment, purified poliovirus RNA (from 500 PFU) was used and some samples were also seeded with BGMK RNA extracted from uninoculated monolayers (Table 3.5). Amplification appeared robust for all samples except those seeded with BGMK RNA and IMS purification (Figure 3.4).

Results for this experiment confirmed the negligible effect of the separate tube RT and Microcon concentration compared to the standard QSD method ($P = 0.10$). It also confirmed that the IMS step was responsible for poor amplification and reduced detection sensitivity. Interestingly, in this experiment there was no difference in the standard QSD amplification of samples seeded with and without BGMK RNA, which is contrary to our previous findings and the impetus for the development of the oligo capture IMS purification approach to begin with. One thought was that differences in batches of pooled BGMK RNA might have an effect on amplification, although monolayers of the same age were used and the same extraction protocol was followed.

Table 3.5
Mean and standard deviation threshold cycle (C_T) values for the biotin oligo capture IMS experiment 3 (500 PFU spike)

Method	Mean ± SD C_T value (n = 5)	
	Molecular grade water only	BGMK RNA added
Standard one-tube reverse transcription QSD	26.10 ± 0.82	26.39 ± 0.28
Separate reverse transcription with biotin-labeled pan-enterovirus primer and Microcon concentration (without IMS)	26.84 ± 0.39	ND*
Separate reverse transcription with biotin-labeled pan-enterovirus primer, Microcon concentration and IMS purification of labeled cDNA	ND*	37.74 ± 1.91

*ND, not done

In a final experiment (Experiment 4), samples were seeded with purified poliovirus RNA and BGMK RNA from two different batches of pooled extracts from replicate uninoculated monolayers (different passage dates). For the standard one-tube QSD, samples seeded with the different batches of BGMK RNA had significantly different amplification between replicates ($P = 0.04$), therefore there was apparently some variability in the assay that was introduced by the background monolayer RNA (Table 3.6). However, the poor recovery of biotin-labeled cDNA by IMS purification was much more substantial.

While oligo capture IMS purified samples seeded with viral RNA (from as low as 100 PFU) and BGMK RNA always tested positive by QSD, C_T values for these samples were consistently much higher than controls. Therefore, our standard one-tube reverse transcription QSD method was used for the remainder of the project.

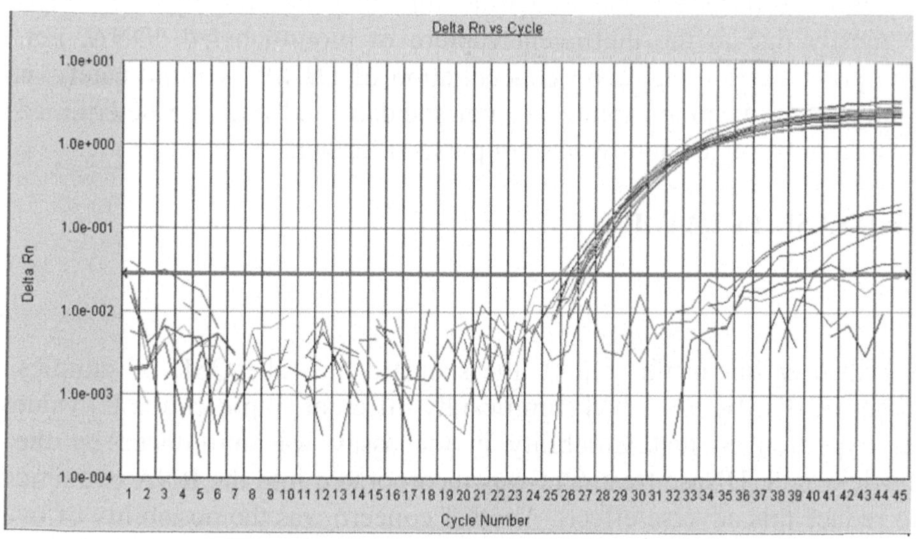

Figure 3.4 QSD amplification plot for the oligo capture IMS experiment 3

Table 3.6
Mean and standard deviation threshold cycle (C_T) values for the biotin oligo capture IMS experiment 4 (100 PFU spike)

Method	Mean ± SD C_T value (n = 5)		
	Molecular grade water only	Batch A BGMK RNA added	Batch B BGMK RNA added
Standard one-tube reverse transcription QSD	26.20 ± 0.15	27.83 ± 1.46	26.21 ± 0.37
Separate reverse transcription with biotin-labeled pan-enterovirus primer, Microcon concentration and IMS purification of labeled cDNA	36.11 ± 0.96	36.76 ± 1.18	36.31 ± 1.10

Conclusions and Recommendations

Oligo capture IMS experiments were performed to evaluate a strategy for the selective separation of viral RNA from BGMK monolayer RNA. Several different conditions were evaluated, including standard one-tube reverse transcription QSD, a separate reverse transcription step, effects of Microcon concentration of cDNA, and the effects of BGMK monolayer RNA. While oligo capture IMS purified samples all tested positive by QSD, C_T values for these samples were consistently higher indicating poor nucleic acid recovery for this purification method. Additional results suggested that the poor recovery of the oligo capture IMS method was mostly due to the inefficient capture of biotin-labeled cDNA, not the separate reverse transcription step or Microcon concentration of cDNA. Unfortunately, conditions for this type of oligo hybridization/capture are not standard and must be determined empirically. Due to time constraints, oligo capture was not pursued further.

EFFECT OF DNASE TREATMENT

Methods

A described earlier in this report, the standard one-tube QSD of samples seeded with different batches of BGMK RNA had significantly different amplification C_T values (P = 0.04). Since it was hypothesized that variability in the assay was introduced by the background monolayer nucleic acid, DNase treatment was incorporated into the RNA extraction protocol in an attempt to reduce this adverse effect. Another concern was the possibility of overloading the RNeasy spin columns or the QSD reactions with too much nucleic acid. Therefore, different volumes of cell monolayer lysate were used to evaluate the RNeasy spin column capacity.

Similarly, different volumes of purified RNA were used to evaluate the effect of total nucleic acid template on QSD amplification.

BGMK and Vero cell lines were cultured as described previously in this report. A factorial design including the following variables was followed for these experiments: 1) RNA with and without DNase, 2) purification of RNA from 560 µl or 140 µl of cell lysate, 3) use of 5 µl or 45 µl of purified RNA as template for QSD.

Results

Across both experiments and for both cell lines the lowest C_T values were obtained by including DNase digestion and using the greatest amount of monolayer lysate (560 µl) and template (45 µl) for QSD. One-way analysis of variance (ANOVA) with Bonferroni multiple pair wise comparisons revealed a significant difference between using 45 µl template of both the 140 µl lysate and the 560 µl lysate ($P = 0.023$). When using 5 µl of template for both the 140 µl and 560 µl lysate, C_T values were also significantly lower for the 560 µl lysate ($P = 0.001$).

BGMK Cells

As seen in Figures 3.5, 3.6, 3.7 and 3.8, BGMK cells inoculated with poliovirus 3 demonstrate significantly lower C_T values when DNase treatment was used in both experiments. The figures present each comparison for purification of partial cell monolayer lysate (140 µl) or the complete lysate (560 µl) and 5 µl or 45 µl (entire sample) purified RNA template per QSD reaction.

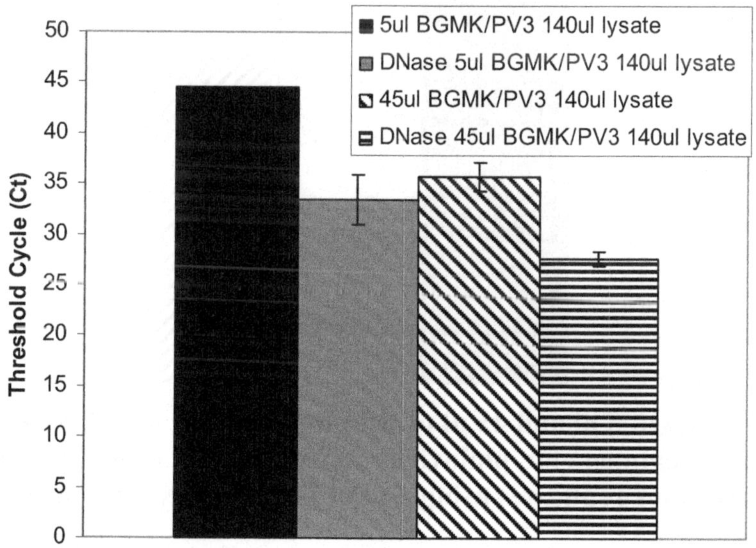

Figure 3.5 Comparison of 5 µl and 45 µl of template and 140 µl of lysate in HCC-QSD with and without DNase treatment for poliovirus 3 and BGMK cells from experiment 1

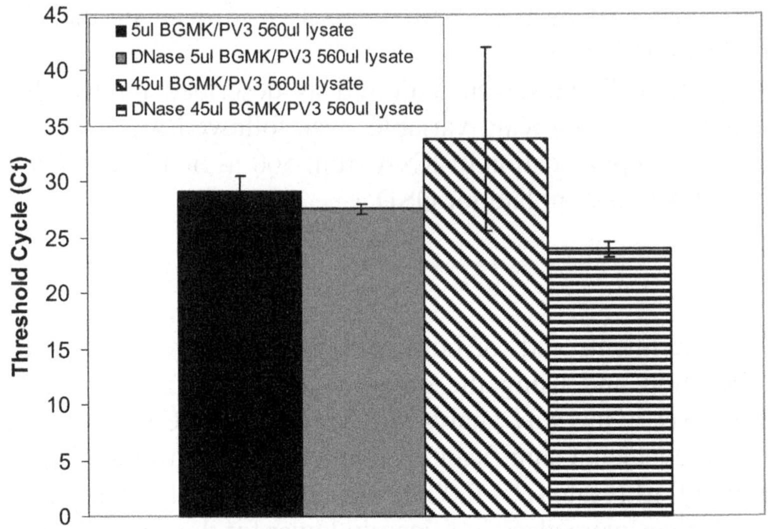

Figure 3.6 Comparison of 5 µl and 45 µl of template and 560 µl of lysate in HCC-QSD with and without DNase treatment for poliovirus 3 and BGMK cells from experiment 1

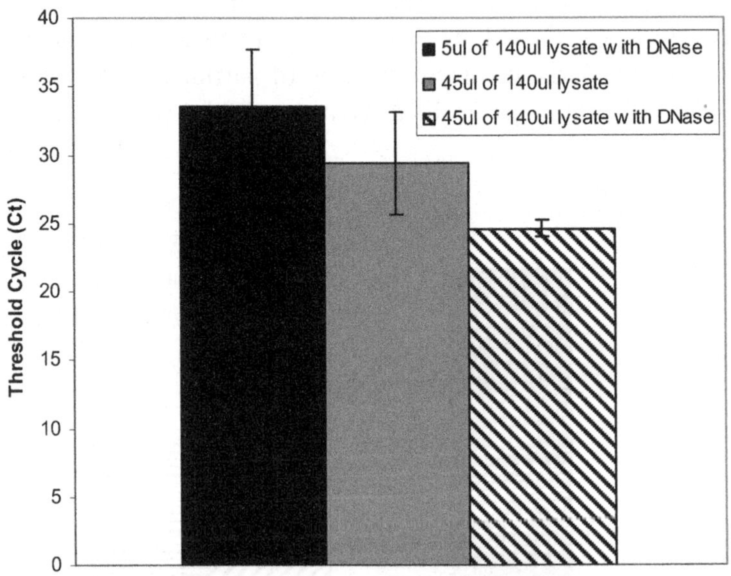

Figure 3.7 Comparison of 5 µl and 45 µl of template and 140 µl of lysate in HCC-QSD with and without DNase treatment for poliovirus 3 and BGMK cells from experiment 2. No amplification was seen for replicates using 5 µl of 140 µl lysate without DNase

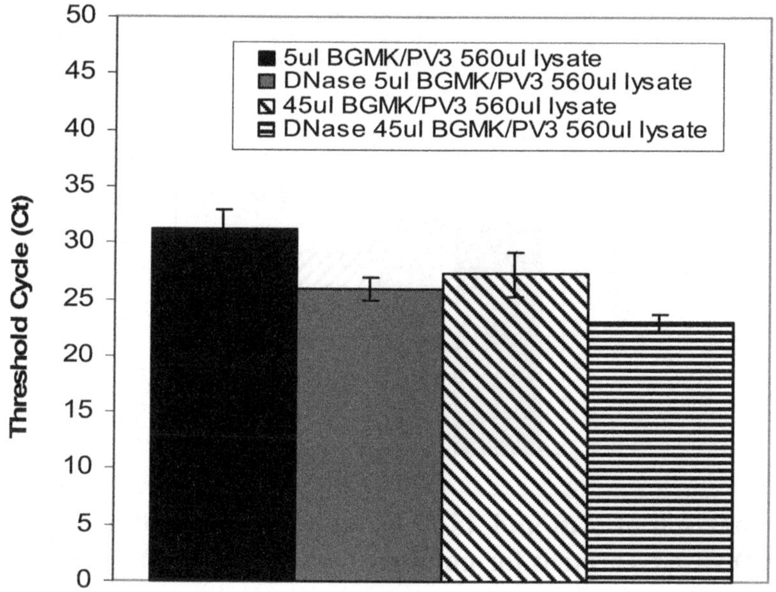

Figure 3.8 Comparison of 5 μl and 45 μl of template and 560 μl of lysate in HCC-QSD with and without DNase treatment for poliovirus 3 and BGMK cells from experiment 2

Vero Cells

A similar comparison was done using Vero cells and poliovirus 3. While the beneficial effect of DNase treatment was not as dramatic as observed for BGMK cells, standard deviations of C_T values for DNase treated samples were much lower than samples without DNase treatment. Figures 3.9, 3.10, 3.11 and 3.12 present the results from both experiments with Vero cells. In most cases the lowest C_T value and smallest standard deviations were obtained by using DNase treatment.

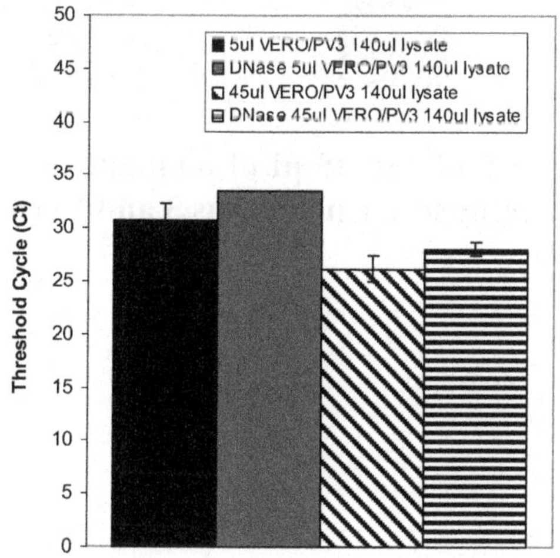

Figure 3.9 Comparison of 5 μl and 45 μl of template and 140 μl of lysate in HCC-QSD with and without DNase treatment for poliovirus 3 and Vero cells from experiment 1

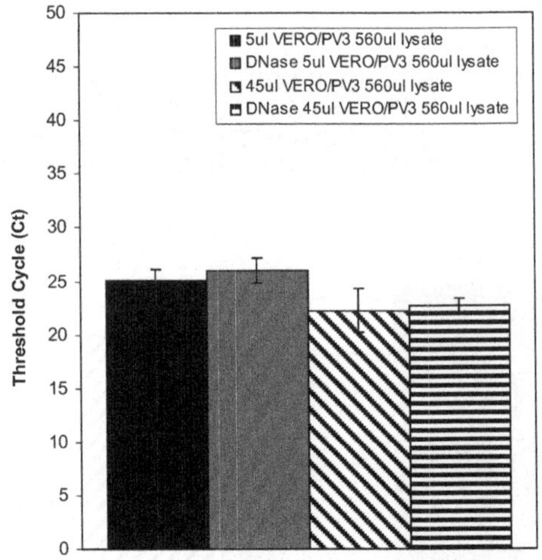

Figure 3.10 Comparison of 5 µl and 45 µl of template and 560 µl of lysate in HCC-QSD with and without DNase treatment for poliovirus 3 and Vero cells from experiment 1

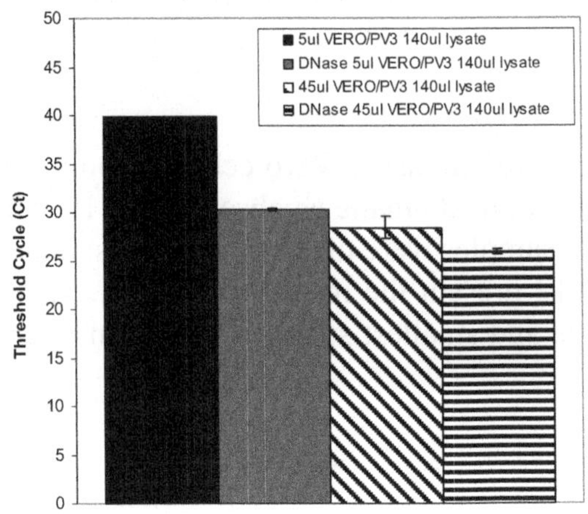

Figure 3.11 Comparison of 5 µl and 45 µl of template and 140 µl of lysate in HCC-QSD with and without DNase treatment for poliovirus 3 and Vero cells from experiment 2

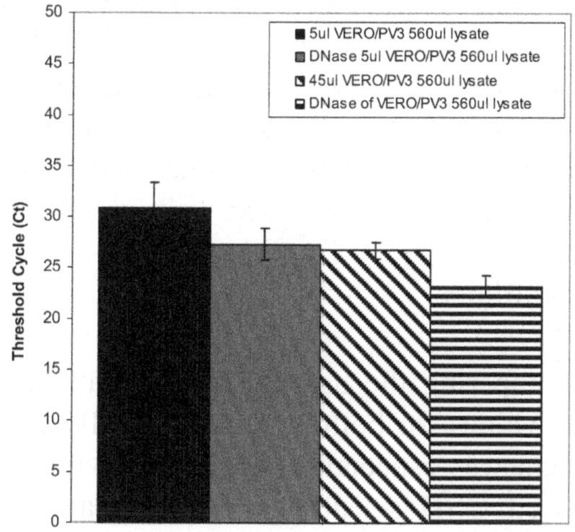

Figure 3.12 Comparison of 5 µl and 45 µl of template and 560 µl of lysate in HCC-QSD with and without DNase treatment for poliovirus 3 and Vero cells from experiment 2

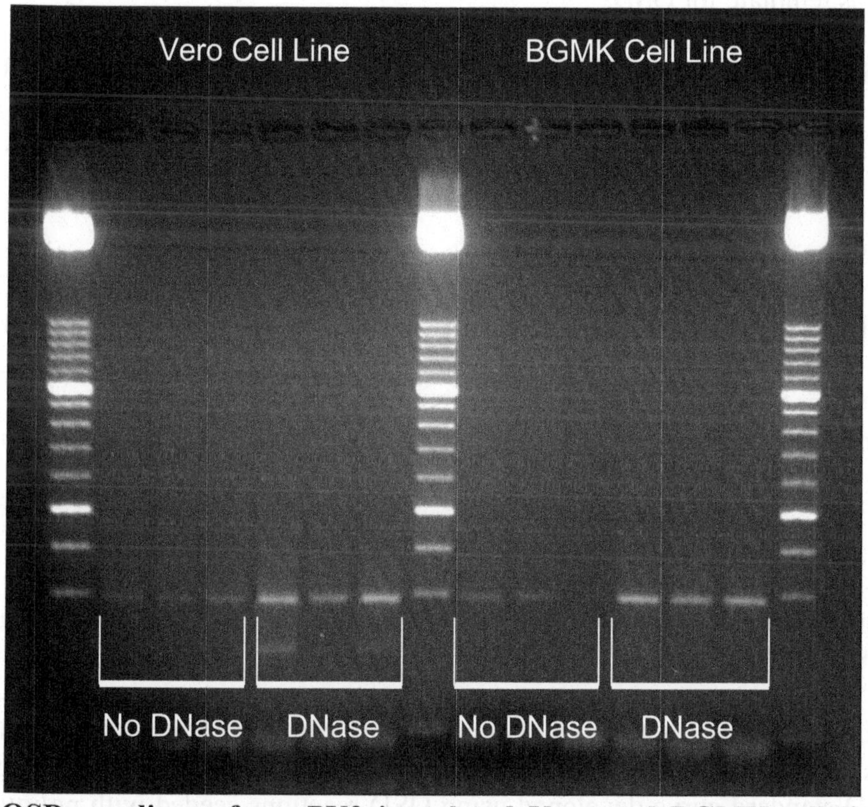

Figure 3.13 QSD amplicons from PV3 inoculated Vero and BGMK cell lines with and without DNase treatment. The third replicate of BGMK cells without DNase had a band which is not visible in the gel image

Gel electrophoresis of QSD amplicons from DNase treated and untreated samples also revealed the beneficial effects of DNase treatment (Figure 3.13). QSD amplicons from samples treated with DNase appear stronger and more distinct than those from samples that did not receive DNase treatment. These results were not surprising given the lower C_T values of DNase treated samples. Importantly, however, is the obvious benefit of DNase treatment even if conventional gel electrophoresis of amplicons is used.

Conclusions and Recommendations

Repeat experiments comparing the use of DNase treatment demonstrated a significant and beneficial decrease in C_T values for both Vero and BGMK cell lines seed with poliovirus. The benefit of DNase treatment was especially pronounced with the BGMK cell line, and was reflected not only in increased QSD sensitivity, but also amplicon quality as detected by conventional gel electrophoresis. Although a less dramatic reduction in C_T values was observed when using DNase with Vero cells line, standard deviations were much tighter and there was also an increase in amplicon quality detected using gel electrophoresis. Both experiments also demonstrated that the best QSD amplification (lowest C_T values) was obtained by using the greatest amount of monolayer lysate (560 µl) for purification of RNA and the entire purified RNA (45 µl) as template for QSD.

HCC-QSD STANDARD CURVES AND EQUATIONS FOR POLIOVIRUS TYPE 3 AND REOVIRUS

Methods

Standard curves of purified poliovirus RNA only and QSD reactions seeded with poliovirus RNA and unseeded BGMK cell monolayer RNA were developed. For the poliovirus only standard curve, the Qiagen Viral RNA kit was used to purify RNA from 0, 0.1, 1, 10, 100, and 1000 PFU of poliovirus 3. For the standard curve using QSD reactions seeded with poliovirus RNA and BGMK cell monolayer RNA (HCC-QSD), the RNA from 0, 1, 10 and 100 PFU poliovirus and BGMK cell monolayer RNA (Qiagen RNeasy kit with DNase treatment) were used.

Results

Standard Curve of Poliovirus RNA and QSD Reactions Seeded With Poliovirus RNA and BGMK RNA

For the poliovirus RNA standard curve, amplification was achieved for samples at each level of virus, although not all replicates of 0.1 PFU per reaction were positive (Figure 3.14). For the HCC-QSD standard curve developed using QSD reactions seeded with poliovirus RNA and BGMK cell RNA (Figure 3.15), detection down to 1 PFU of poliovirus RNA in the presence of BGMK RNA from 12-well monolayers was achieved. This disparity was likely due to Poisson distribution of low concentration template RNA. This possible explanation is supported by the finding that C_T values were not significantly different ($P > 0.05$) between reactions seeded with only poliovirus RNA compared to reactions seeded with both poliovirus and BGMK cell monolayer RNA (Table 3.7). Standard curves and linear equations are presented in Figures 3.16 and 3.17.

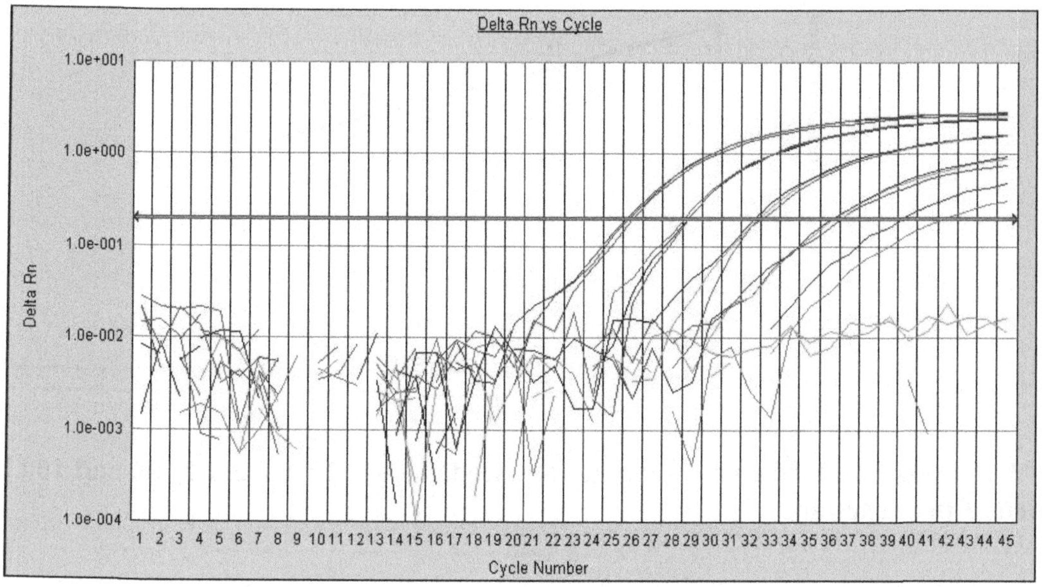

Figure 3.14 QSD standard curve for RNA from 0.1, 1, 10, 100 and 1000 PFU of poliovirus 3 per reaction

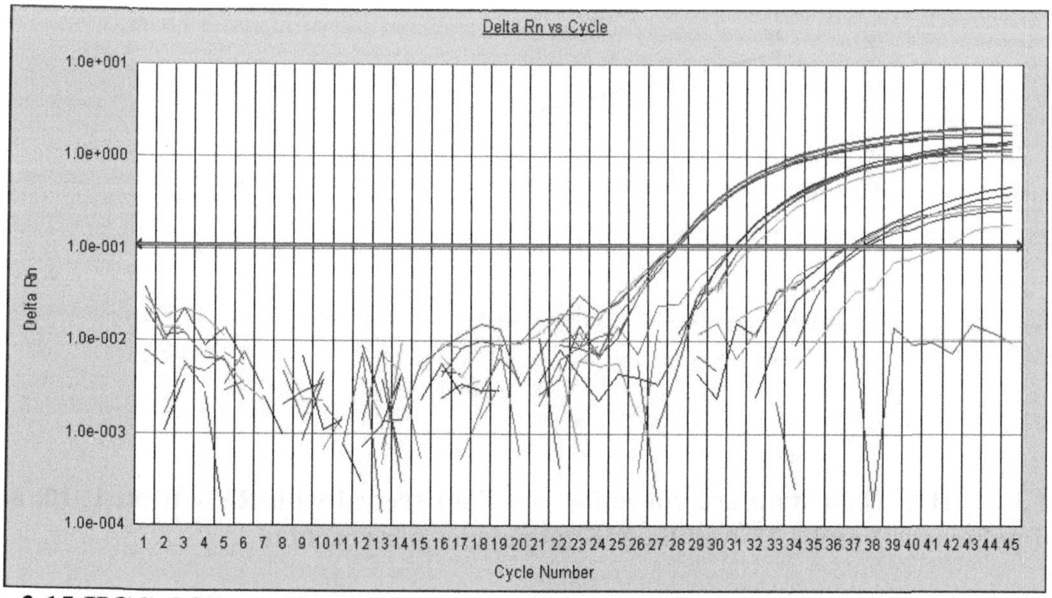

Figure 3.15 HCC-QSD standard curve for reactions seeded with RNA from 1, 10, and 100 PFU of poliovirus 3 and a 12-well BGMK monolayer per reaction

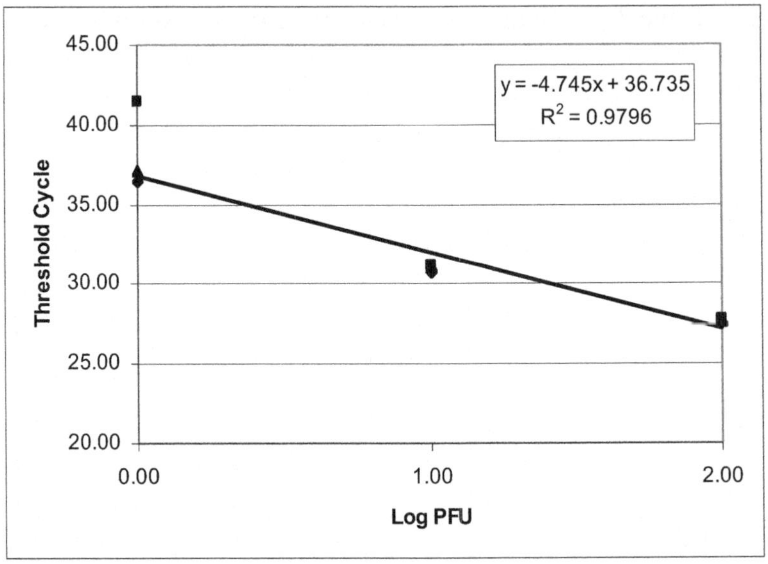

Figure 3.16 QSD standard curve for reactions seeded with RNA from 1, 10, and 100 PFU of poliovirus 3 per reaction

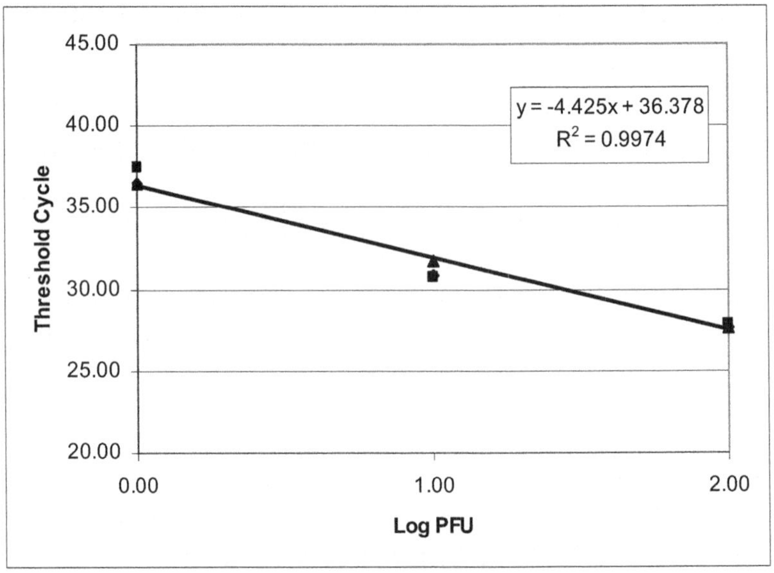

Figure 3.17 HCC-QSD standard curve for reactions seeded with RNA from 1, 10, and 100 PFU of poliovirus 3 and a 12-well BGMK monolayer per reaction

Table 3.7
Comparison of C_T values for QSD reactions seeded with poliovirus RNA and BGMK cell monolayer RNA

Poliovirus RNA Seed Level	Mean ± SD C_T Value	
	Virus RNA Only	Virus RNA and BGMK RNA
100 PFU	27.62 ± 0.10	27.74 ± 0.13
10 PFU	31.04 ± 0.27	31.07 ± 0.54
1 PFU	38.34 ± 2.69	36.84 ± 0.59

Standard Curve of Reovirus RNA

The standard curve for reovirus was created using purified reovirus RNA dilutions 10^{-1} through 10^{-5}. Since the actual copy number of reovirus genomes was unknown, it was estimated as 10,000 copies for the 10^{-1} dilution down to one copy for the 10^{-5} dilution based on the 50% tissue culture infectious dose ($TCID_{50}$) provided by ATCC. The reovirus standard curve and linear equation are presented in Figure 3.18.

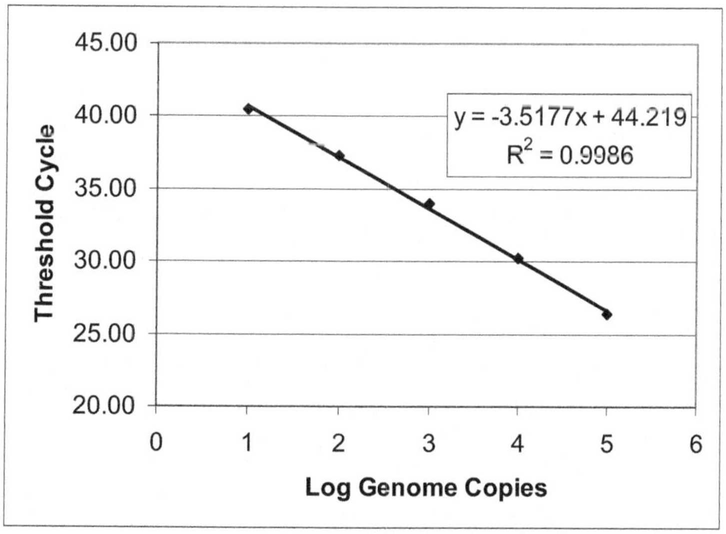

Figure 3.18 HCC-QSD standard curve for reactions seeded with reovirus RNA 10^{-1} through 10^{-5} dilutions (approximately 10,000 to 1 genome copies)

Conclusions and Recommendations

HCC-QSD standard curves and equations were successfully created for both poliovirus and reovirus. The poliovirus standard curve equation was used to calculate virus inactivation with chlorine and ultraviolet light (see Chapter 4).

Chapter 4
EVALUATION of HCC-QSD to Distinguish Potentially Infectious Viruses From Chlorine- and UV- INACTIVATED Viruses

INTRODUCTION

Chlorine typically inactivates viruses by altering the capsid, which causes a structural change that inhibits viral adsorption to host cells. Chlorine may also cause some damage to the viral genome. Results from a poliovirus inactivation study show that changes in virus structure are dependent on the initial chlorine concentration applied, but changes are not the initial cause of inactivation. Higher concentrations of chlorine result in a structural effect while lower doses of chlorine show an effect in the RNA (Alvarez and O'Brien 1982). Therefore, the mechanism of chlorine inactivation of viruses can vary depending on the concentrations of chlorine applied. In contrast, the mechanism of viral inactivation by ultraviolet light is through damage to the viral genome.

Cell culture has been used with PCR (ICC-PCR) to measure the inactivation of poliovirus using chlorine (Blackmer, Reynolds et al. 2000). However, the ICC-PCR approach used differs from HCC-QSD since ICC-PCR relies on replication of the virus in cell culture. Blackmer, et al. (Blackmer, Reynolds et al. 2000) demonstrated the rapid and effective use of integrated cell culture-PCR to detect the infectivity of poliovirus after chlorine inactivation, and concluded that 0.5 mg of free chlorine per liter for 10 minutes (5 mg·min L^{-1} CT; product of concentration in mg per liter and time in minutes) was required for complete inactivation. Previous work investigating the use of QSD for measuring direct UV damage to *Cryptosporidium* oocysts (without cell culture) has demonstrated enough DNA damage that there is no longer any amplification when using 40 to 60 mJ cm^{-2} (Di Giovanni and Ronquillo 2002).

Preliminary evidence suggested that the HCC method may be capable of distinguishing between potentially infectious viruses and those that have been inactivated by chlorine (Di Giovanni, Karim et al. 2002). To further the investigate the potential for HCC-QSD to differentiate inactivated from infectious viruses, poliovirus was exposed to zero, low, medium and high free chlorine CTs; and zero, medium, low and high doses of low-pressure monochromatic 254 nm ultraviolet light. Free chlorine oxidant demand of the samples was taken into consideration and experiments were performed at pH 7.0 and 25 °C. Samples were analyzed by both conventional BGMK plaque assay and HCC-QSD.

CHLORINE INACTIVATION

Methods

Virus Purification

Polyethylene (PEG) glycol precipitation was used in order to further purify the poliovirus from cell lysates (Thurston-Enriquez, Haas et al. 2003). To a total volume of 30 ml virus suspension, 2.7 g of PEG and 1.74 g of NaCl were added and the suspension mixed. The suspension was stirred overnight at 4°C and then centrifuged at 10,000 X g at 4°C for 30 minutes to remove cell debris. The virus pellet was resuspended in 10 ml buffered demand free water (BDF; pH 7.0). BDF water was prepared by dissolving 0.54 g of Na_2HPO_4 (anhydrous) and 0.88g of KH_2PO_4 (anhydrous) per liter of deionized water with pH adjusted to 7.0.

Virus Dispersion

Dispersion of viral stocks was accomplished by chloroform extraction (Thurston-Enriquez, Haas et al. 2003). The virus stock was combined with an equal volume of chloroform and vortexed continuously for 10 minutes. Once homogenized, the virus suspension was centrifuged at 10,000 X g at 4°C for 15 minutes. Dispersed viral stocks were stored at 4°C until use.

Free Chlorine Solution and Measurements

The N,N-Diethyl-p-phenylenediamine (DPD) spectrophotometric method (Standard Methods, 1995) was used to measure and create a free chlorine standard curve using potassium permanganate standards. Working stock chlorine solution was prepared by dilution of 5% sodium hypochlorite solution (5% available chlorine) in BDF to a final concentration of 1 part per million (ppm) free chlorine. The solution was protected from light and used immediately after preparation. A ten-fold dilution of the 1 ppm stock (approximately equal to 0.1 ppm) was used for lower CT exposures. The ability to neutralize the free chlorine solution was verified using a 10% sodium thiosulfate solution.

Determining Oxidant demand of a sample

Free chlorine residual was tested at each CT value to ensure chlorine residual measurements remained sufficient. Total and free chlorine levels were measured and recorded to calculate actual chlorine exposure using the DPD spectrophotometric method and Hach water quality test strips. Virus/chlorine suspension was prepared using 1000 PFU per 100 μl of poliovirus type 3 and chlorine solution. The virus/chlorine suspension was measured for chlorine residual after 1 min, 5 min, and 15 min. Oxidant demand was determined to be 4 ppm free chlorine, which was taken into account when calculating the contact times for each CT level of chlorine (0, 1, 10, 50, 100 mg·min L^{-1}). All contact times had measurable residual free chlorine of at least 1 ppm.

Enterovirus Plaque Assay

Four to five day old confluent BGMK monolayers in 6-well plates were used for plaque assays. For each level of disinfectant, three 100 μl replicates of undiluted, 10^{-1} and 10^{-2} dilutions of virus suspension were assayed to ensure a countable number of plaques. Plates were incubated in a 37°C CO_2 incubator for 2 hours, rocking every 15 minutes to evenly distribute samples across wells and prevent the monolayers from drying out. Overlay agar was prepared with 100 ml of 2X MEM, 100 ml of 1.5% agar, 4 ml FBS, 6 ml HEPES, 2 ml sodium bicarbonate, 2 ml glutamine (200 mM), 2 ml Fungizone, 2 ml Pen/Strep, 2 ml Kanamycin (100X or 10 mg/ml). After the 2-hour incubation, 2 ml of overlay agar was added to each of the wells without washing the monolayers and plates incubated at 37°C CO_2 for 36 - 40 hours. After removing the overlay agar each well was stained with 0.01% crystal violet and plaques were enumerated and recorded.

HCC-QSD of Inactivated samples

After the disinfection procedures, viral suspensions were used to inoculate BGMK cell monolayers and a two-hour host cell capture time was used. RNA extractions were carried out using the Qiagen RNeasy protocol with DNase treatment as described in Chapter 3. Samples that did not amplify were given a C_T value of 45, the maximum number of cycles performed for HCC-QSD. The poliovirus HCC-QSD standard curve described in Chapter 3 was used to calculate the equivalent poliovirus PFU from HCC-QSD C_T values.

Results

Both HCC-QSD and conventional plaque assay clearly demonstrated free-chlorine inactivation of the poliovirus and yielded very similar results. Both HCC-QSD data and plaque assay demonstrated a complete inactivation of poliovirus for 0.5 CT and 1 CT for both repeat experiments (Figures 4.1 and 4.2). There were no significant differences ($P > 0.05$) between 0, 0.01, or 0.1 CT detected using either plaque assay or HCC-QSD.

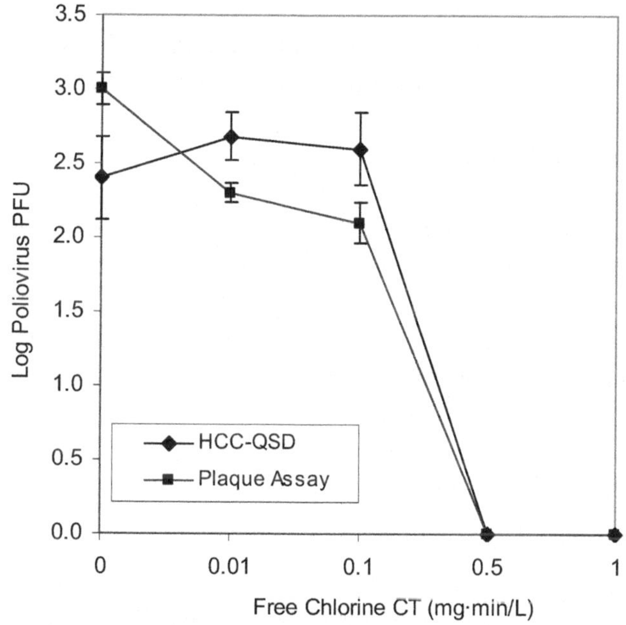

Figure 4.1 Free chlorine inactivation of poliovirus type 3 measured using HCC-QSD (diamonds) and conventional BGMK plaque assay (squares) for experiment 1

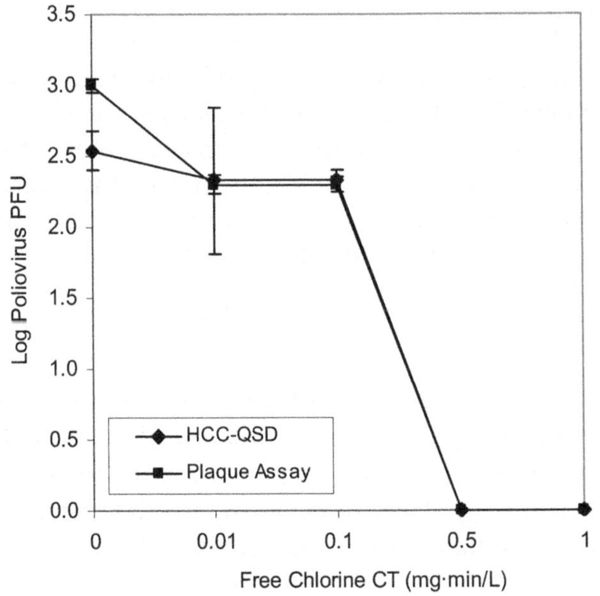

Figure 4.2 Free chlorine inactivation of poliovirus type 3 measured using HCC-QSD (diamonds) and conventional BGMK plaque assay (squares) for experiment 2

Conclusions and Recommendations

Results of the free-chlorine inactivation experiments clearly demonstrated that HCC-QSD is capable of distinguishing between infectious and chlorine-inactivated poliovirus and can provide results similar to the conventional plaque assay. Both methods showed complete inactivation of the poliovirus at 0.5 CT, with no significant inactivation at lower CTs detectable using either method. Based on these results, HCC-QSD appears to be a rapid alternative to the plaque assay for quantifying infectious poliovirus.

ULTRAVIOLET LIGHT (UV) INACTIVATION

Methods

Virus suspensions were made using 1000 PFU poliovirus type 3 vaccine strain diluted in reagent grade water. Suspensions were put in small Petri dishes (60 x 15mm) and exposed to zero, medium, low and high doses (0, 10, 40, and 100 mJ cm^{-2}) of ultraviolet light (UV) using a bench-scale monochromatic 254 nm low-pressure UV system. This system consisted of a Stratalinker 1800 UV crosslinker (Stratagene, La Jolla, CA) with integrated calibrated radiometer. UV-exposed and control viral suspensions were assayed by the BGMK plaque assay and HCC-QSD as described for the free-chlorine experiments above.

Results

Three UV inactivation experiments were performed in order to evaluate reproducibility of results and confirm findings. Mean log_{10} PFU values and standard deviations of samples for all three experiments are presented in Table 4.1. In all three experiments, the plaque assay results revealed a clear inactivation trend, with increasing inactivation observed as UV exposure increased and complete inactivation at ≥ 40 mJ cm^{-2} (Table 4.1 and Figures 4.3, 4.4, and 4.5). In contrast, HCC-QSD detected only a minor decrease in infectious poliovirus at this same dose for two of three experiments (Table 4.1 and Figures 4.3, 4.4, and 4.5).

Plaque assay results for all three experiments indicated that there were significant differences ($P < 0.05$) between 0 mJ cm^{-2} and the 10, 40, 100 mJ cm^{-2} UV treatment levels. In contrast, statistical analysis of HCC-QSD data revealed no significant differences ($P > 0.05$) for any of the UV levels in experiment 1, and only significant differences between the unexposed controls and the 40 and 100 mJ cm^{-2} doses for experiments 2 and 3 ($P = 0.033$ and $P = 0.006$; $P = 0.053$, $P = 0.000$; respectively). Although statistically significant, the levels of inactivation at these doses determined using HCC-QSD were clearly underestimates when compared to the plaque assay results.

Table 4.1
Mean and standard deviation of poliovirus 3 (Log_{10} PFU) for ultraviolet light inactivation experiments

Ultraviolet Light Inactivation (mJ cm^{-2})	Mean ± SD Log_{10} PFU value					
	Exp 1		Exp 2		Exp 3	
	Plaque Assay	HCC-QSD	Plaque Assay	HCC-QSD	Plaque Assay	HCC-QSD
0	3.00 ± 0.10[a]	2.72 ± 0.21[a]	3.07 ± 0.06[a]	2.93 ± 0.07[a]	3.03 ± 0.06[a]	2.66 ± 0.16[a]
10	1.67 ± 0.15[b]	2.65 ± 0.31[a]	1.53 ± 0.06[b]	2.85 ± 0.13[b]	1.60 ± 0.10[a]	2.54 ± 0.38[a]
40	0.00 ± 0.00[c]	2.77 ± 0.10[a]	0.00 ± 0.00[c]	2.66 ± 0.20[c]	-0.50 ± 0.00[c]	0.84 ± 1.50[b]
100	0.00 ± 0.00[c]	2.49 ± 0.14[a]	0.00 ± 0.00[c]	2.59 ± 0.17[c]	0.00 ± 0.00[c]	-1.12 ± 1.29[c]

[*]Groups within each column with different letters are significantly different ($P < 0.05$)

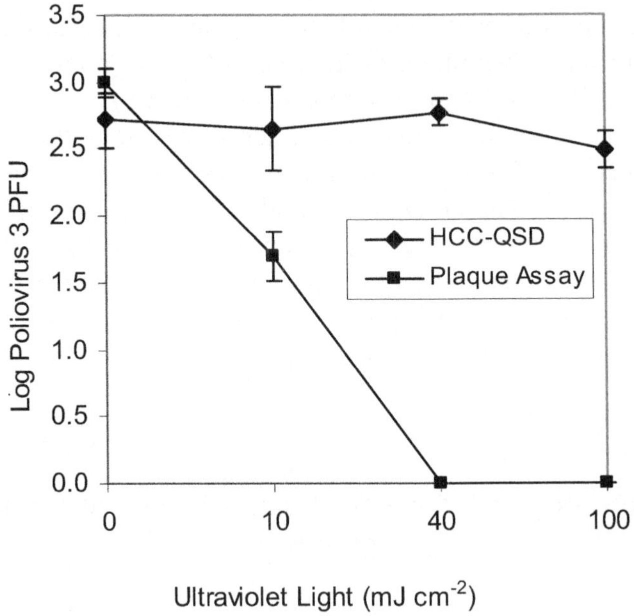

Figure 4.3 Ultraviolet light inactivation of poliovirus type 3 comparison using HCC-QSD (diamonds) and plaque assay (squares) for experiment 1

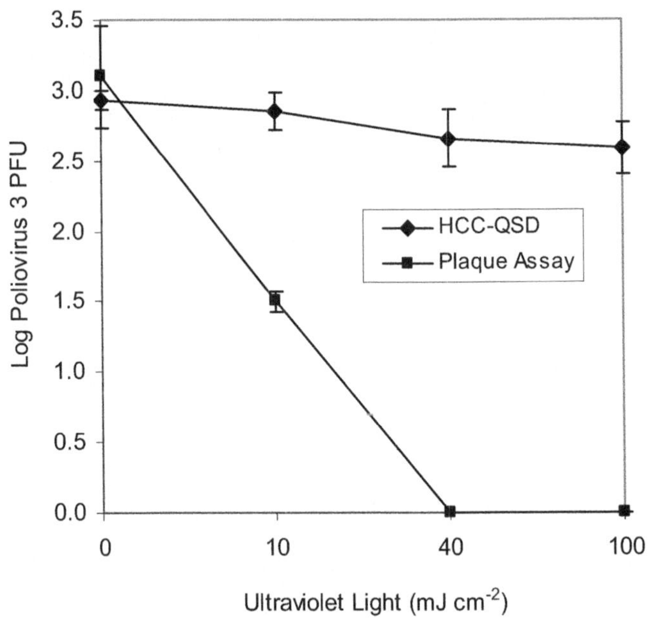

Figure 4.4 Ultraviolet light inactivation of poliovirus type 3 comparison using HCC-QSD (diamonds) and plaque assay (squares) for experiment 2

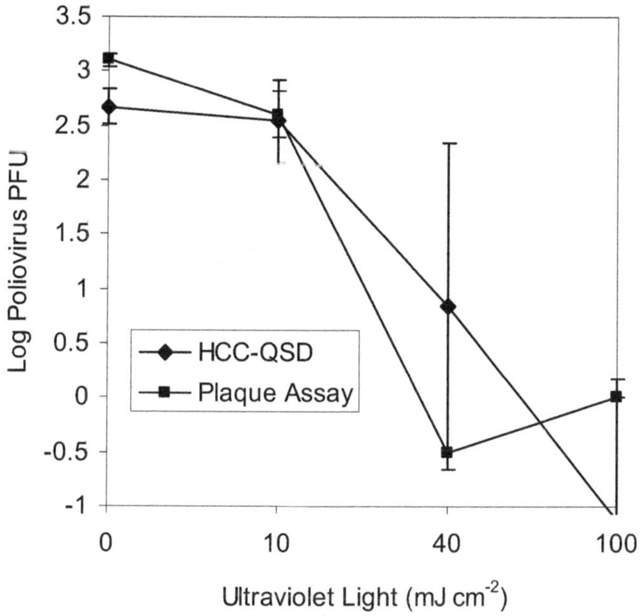

Figure 4.5 Ultraviolet light inactivation of poliovirus type 3 comparison using HCC-QSD (diamonds) and plaque assay (squares) for experiment 3

Conclusions and Recommendations

Plaque assay results for all experiments consistently revealed virus inactivation with a UV dose as small as 10 mJ cm^{-2}, with complete inactivation at 40 mJ cm^{-2} or higher. In contrast to the plaque assay, HCC-QSD data for only two of three experiments revealed a significant difference between 40 mJ cm^{-2} and 100 mJ cm^{-2} UV doses compared to the untreated control. Although statistically significant, the levels of inactivation at these doses determined using HCC-QSD were clearly underestimates when compared to the plaque assay results. HCC-QSD results were likely a reflection of UV damage to the viral genome, rather than an effect on the viral protein structure (i.e. viruses still adsorbed to monolayers), posing a challenge for the HCC-QSD method. Currently, we do not recommend the use of HCC-QSD for poliovirus UV-inactivation work.

CHAPTER 5
SUMMARY AND CONCLUSIONS

Waterborne viruses present a significant threat to human health, especially for the growing immunocompromised population. Currently, routine water monitoring and risk assessment for infectious viruses in water is hampered by the lack of rapid, specific and cost-effective methods. In this study, significant progress was made in the further development of HCC-QSD as a rapid method for the detection and quantitation of potentially infectious viruses in water. Specific conclusions of the project are:

1. Although the BGMK and Vero cell lines are best suited for the propagation of poliovirus and reovirus, respectively, either cell line could be used to capture the viruses. This provides some evidence for the possible use of a single cell monolayer for HCC-QSD to detect different viruses in the same sample. However, the capture dynamics were different between the two cell lines and viruses, indicating that further optimization would be required for this approach.
2. Monolayers grown in 12-well plates and inoculated with 100 μl sample volumes proved to be effective for the virus capture step.
3. The preparation of inocula in serum-free Minimum Essential Medium (MEM) is recommended over the use of phosphate buffered saline (PBS). The use of MEM helped maintain cell monolayer integrity compared to PBS, especially for the BGMK cell line.
4. Comparison of the two Qiagen RNA extraction kits revealed that the RNeasy kit without the use of QIAshredders or β-mercaptoethanol provided the most reliable extraction of viral RNA from cell monolayers.
5. Oligo capture immunomagnetic separation (IMS) was capable of separating viral RNA from host cell monolayer nucleic acids. However, RNA recovery efficiency was unacceptably low and further optimization of oligo capture conditions are needed prior to using this purification approach.
6. Addition of DNase treatment to the Qiagen RNeasy extraction protocol was shown to have a significant and beneficial effect on QSD amplification. The benefit of DNase treatment was especially pronounced with the BGMK cell line, and was reflected not only in increased QSD sensitivity, but also amplicon quality as detected by conventional gel electrophoresis.
7. In contrast to the conventional plaque assay, HCC-QSD was only capable of detecting UV-inactivation of poliovirus at doses of $\geq$ 40 mJ cm^{-2}. The greater sensitivity of the plaque assay compared to HCC-QSD was likely due to the inability of UV-inactivated virus to replicate in cell culture, whereas significant UV-induced damage to the viral genome was required before a decrease in HCC-QSD signal was observed. Therefore, HCC-QSD is currently not recommended for UV-inactivation work.
8. There was excellent agreement between HCC-QSD and the conventional plaque assay for the detection of infectious and free-chlorine inactivated virus. In repeat experiments both HCC-QSD and the conventional plaque assay detected complete inactivation of poliovirus with 0.5 mgmin L^{-1} free chlorine concentration × time (CT). This suggests that the HCC-QSD assay may be a useful alternative to the plaque assay for the rapid monitoring of chlorine-disinfected water.

FUTURE RESEARCH

Further development of the HCC-QSD method is needed prior to routine field application. In addition to poliovirus and reovirus, we had intended to work with hepatitis A and adenovirus 40 and 41 for this project. However, we were unsuccessful in propagating these viruses. Therefore, these viruses as well as others should be evaluated for use in HCC-QSD. Chlorine disinfection studies with these other viruses should be performed to understand the strengths and limitations of HCC-QSD.

Since HCC-QSD is potentially capable of recovering a variety of enteric viruses from a single sample, multiplex QSD primers and probes are needed to allow simultaneous detection of different viruses. With the current availability of novel PCR reagents to overcome some of the recognized problems with multiplexing, the use of previously developed primers and probes that are compatible for multiplexing may avoid the need to develop new primers and probes.

Evaluation of HCC-QSD with environmental water sample matrices is needed to integrate upstream sample processing. Of the currently available methods for the concentration of viruses in water samples, the glass wool filter method (Vilaginès, Sarrette et al. 1997; Grabow, Taylor et al. 2001) followed by centrifugation or precipitation of viruses appears particularly promising. This method of virus concentration is efficient, inexpensive, and capable of concentrating large volumes of water (e.g. 1500 L drinking water) down to a small volume (approximately 2 ml) suitable for HCC-QSD.

REFERENCES

Abbaszadegan, M., P. Stewart, et al. (1999). "A strategy for detection of viruses in groundwater by PCR." Appl Environ Microbiol **65**(2): 444-9.

Alvarez, M. E. and R. T. O'Brien (1982). "Effects of chlorine concentration on the structure of poliovirus." Appl Environ Microbiol **43**(1): 237-9.

Blackburn, B. G., G. F. Craun, et al. (2004). "Surveillance for waterborne-disease outbreaks associated with drinking water--United States, 2001-2002." MMWR Surveill Summ **53**(8): 23-45.

Blackmer, F., K. A. Reynolds, et al. (2000). "Use of integrated cell culture-PCR to evaluate the effectiveness of poliovirus inactivation by chlorine." Appl Environ Microbiol **66**(5): 2267-8.

Chapron, C. D., N. A. Ballester, et al. (2000). "Detection of astroviruses, enteroviruses, and adenovirus types 40 and 41 in surface waters collected and evaluated by the information collection rule and an integrated cell culture-nested PCR procedure." Appl Environ Microbiol **66**(6): 2520-5.

Craun, G. F. and R. L. Calderon (1996). Microbial risks in groundwater systems: epidemiology of waterborne outbreaks. Under the Microscope, Examining Microbes in Groundwater. Denver, Colorado, American Water Works Association: 9-20.

Di Giovanni, G. D., M. R. Karim, et al. (2002). Overcoming Molecular Sample Processing Limitations: Quantitative PCR. Alexandria, VA, Water Environment Research Foundation and International Water Association.

Di Giovanni, G. D. and N. Ronquillo (2002). Direct molecular quantitation of UV-induced damage to *Cryptosporidium* oocysts. Proceedings of the AWWA Water Quality Technology Conference, American Water Works Association, Denver, Colo. [on CD-ROM].

Enriquez, C. E., C. J. Hurst, et al. (1995). "Survival of the enteric adenoviruses 40 and 41 in tap, sea and wastewater." Water Res **29**: 2548-2553.

Gilgen, M., B. Wegmuller, et al. (1995). "Reverse transcription PCR to detect enteroviruses in surface water." Appl Environ Microbiol **61**(4): 1226-31.

Grabow, W. O., M. B. Taylor, et al. (2001). "New methods for the detection of viruses: call for review of drinking water quality guidelines." Water Sci Technol **43**(12): 1-8.

Greening, G. E., J. Hewitt, et al. (2002). "Evaluation of integrated cell culture-PCR (C-PCR) for virological analysis of environmental samples." J Appl Microbiol **93**(5): 745-50.

Lee, S. H. and S. J. Kim (2002). "Detection of infectious enteroviruses and adenoviruses in tap water in urban areas in Korea." Water Res **36**(1): 248-56.

Matsuura, K., S. Hasegawa, et al. (1984). "Viral pollution of the rivers in Toyama City." Microbiol Immunol **28**(5): 575-88.

Monpoeho, S., M. Coste-Burel, et al. (2002). "Application of a real-time polymerase chain reaction with internal positive control for detection and quantification of enterovirus in cerebrospinal fluid." Eur J Clin Microbiol Infect Dis **21**(7): 532-6.

Monpoeho, S., A. Dehee, et al. (2000). "Quantification of enterovirus RNA in sludge samples using single tube real-time RT-PCR." BioTechniques **29**: 88-93.

Muscillo, M., A. Carducci, et al. (1997). "Enteric virus detection in Adriatic seawater by cell culture, polymerase chain reaction, and polyacrylamide gel electrophoresis." Water Res **31**: 1980-1984.

Muscillo, M., G. La Rosa, et al. (2001). "A new RT-PCR method for the identification of reoviruses in seawater samples." Water Res **35**(2): 548-556.

Reynolds, K. A. (2004). "Integrated cell culture/PCR for detection of enteric viruses in environmental samples." Methods Mol Biol **268**: 69-78.

Schwab, K. J., R. De Leon, et al. (1995). "Concentration and purification of beef extract mock eluates from water samples for the detection of enteroviruses, hepatitis A virus, and Norwalk virus by reverse transcription-PCR." Appl Environ Microbiol **61**(2): 531-7.

Shields, A. F., R. C. Hackman, et al. (1985). "Adenovirus infections in patients undergoing bone-marrow transplantation." N Engl J Med **312**(9): 529-33.

Spinner, M. L. and G. D. Di Giovanni (2001). "Detection and identification of mammalian reoviruses in surface water by combined cell culture and reverse transcription-PCR." Appl Environ Microbiol **67**(7): 3016-3020.

Thurston-Enriquez, J. A., C. N. Haas, et al. (2003). "Chlorine inactivation of adenovirus type 40 and feline calicivirus." Appl Environ Microbiol **69**(7): 3979-3985.

Uchiyama, A. and D. G. Besselsen (2003). "Detection of Reovirus type 3 by use of fluorogenic nuclease reverse transcriptase polymerase chain reaction." Lab Anim **37**(4): 352-9.

USEPA (2003). Workshop to Develop a Protocol for Reliable Genetic Methods for the Detection of Viruses for Use in EPA's Water Program Report, Office of Ground Water and Drinking Water, Cincinatti, Ohio: 16.

Vilaginès, P., B. Sarrette, et al. (1997). "Round robin investigation of glass wool method for poliovirus recovery from drinking water and sea water." Water Sci Technol **35**(11-12): 445-449.

Zahradnik, J. M., M. J. Spencer, et al. (1980). "Adenovirus infection in the immunocompromised patient." Am J Med **68**(5): 725-32.

ABBREVIATIONS

ANOVA	analysis of variance
ATCC	American Type Culture Collection
BGMK	Buffalo Green monkey kidney cells
cDNA	complementary deoxyribonucleic acid
CO_2	carbon dioxide
CT	the product of concentration and time
C_T	threshold cycle
DNA	deoxyribonucleic acid
DNase	deoxyribonuclease
FBS	fetal bovine serum
h	hour(s)
HCC	host cell capture
HCC-QSD	host cell capture quantitative sequence detection
HEPES	hydroxymethylpiperazine ethanesulfonic acid
ICC-PCR	integrated cell culture-polymerase chain reaction
IMS	immunomagnetic separation
MEM	Minimal Essential Medium
mg	milligram(s)
mg·min/L	milligrams minute per liter
min	minute(s)
$mJ\ cm^2$	millijoules per centimeter squared
ml	milliliter(s)
mM	millimolar
µl	microliter(s)
NTC	no template control
PBS	phosphate buffered saline
PCR	polymerase chain reaction
PFU	plaque-forming unit
ppm	parts per million
PV3	poliovirus 3
QSD	quantitative sequence detection
RNA	ribonucleic acid
RNase	ribonuclease

RT-PCR	reverse transcription-polymerase chain reaction
sec	second(s)
Taq	*Thermus aquaticus*
UV	ultraviolet light

IWA Publishing's authorised EU representative for General Product Safety Regulations is Diane D'Arras, 15 rue Duret, 75116 Paris, France, e-mail: safety@iwap.co.uk.

Printed and bound by CPI Group (UK) Ltd, Croydon, CR0 4YY
17/09/2026
02218811-0004